WILD
FIRE

WILD FIRE

DISPATCHES FROM A COUNTRY ABLAZE

JESSE WINTER

HarperCollins*Publishers*Ltd

Wild Fire
Copyright © 2026 by Jesse Winter.
All rights reserved.

Published by HarperCollins Publishers Ltd

FIRST EDITION

No part of this book may be used or reproduced in any manner whatsoever without written permission.

Without limiting the exclusive rights of any author, contributor or the publisher of this publication, any unauthorized use of this publication to train generative artificial intelligence (AI) technologies is expressly prohibited. HarperCollins also exercise their rights under Article 4(3) of the Digital Single Market Directive 2019/790 and expressly reserve this publication from the text and data mining exception.

HarperCollins books may be purchased for educational, business, or sales promotional use through our Special Markets Department.

HarperCollins Publishers Ltd
Bay Adelaide Centre, East Tower
22 Adelaide Street West, 41st Floor
Toronto, Ontario, Canada
M5H 4E3

www.harpercollins.ca

HarperCollins Publishers
Macken House, 39/40 Mayor Street Upper
Dublin 1, D01 C9W8, Ireland
ww.harpercollins.com

Library and Archives Canada Cataloguing in Publication

Title: Wild fire : dispatches from a country ablaze / Jesse Winter.
Names: Winter, Jesse, author.
Description: Includes bibliographical references and index.
Identifiers: Canadiana (print) 20260120553 | Canadiana (ebook) 20260120561 | ISBN 9781443475822 (hardcover) | ISBN 9781443475839 (EPUB)
Subjects: LCSH: Wildfires—Canada. | LCSH: Forest fire fighters—Canada. | LCSH: Volunteer fire fighters—Canada.
Classification: LCC SD421.34.C3 W56 2026 | DDC 363.37/90971—dc23

Printed and bound in the United States of America

26 27 28 29 30 LBC 5 4 3 2 1

For the thousands of firefighters, pilots, drivers and contractors who risk their lives to help keep us safe.

CONTENTS

AUTHOR'S NOTE IX

INTRODUCTION 1

CHAPTER 1: **K21620** 15

CHAPTER 2: **IGNITION** 37

CHAPTER 3: **TRAPPED** 59

CHAPTER 4: **CHAOS, COURAGE AND CONSPIRACY** 85

CHAPTER 5: **RADIO SILENCE** 111

CHAPTER 6: **EVACUATION** 133

CHAPTER 7: **CITIES ON FIRE** 157

CHAPTER 8: **CALL FOR AID** 177

CHAPTER 9: **WE'RE NOT GOING HOME TODAY** 193

CHAPTER 10: **THE BATTLE FOR JASPER** 205

CHAPTER 11: **RESET** 229

CHAPTER 12: **COMMUNITIES AT THE READY** 255

EPILOGUE: **OUR FORESTS AND OUR FIRES ARE CHANGING** 279

ACKNOWLEDGEMENTS 289

ENDNOTES 293

AUTHOR'S NOTE

When I first stepped onto a wildland fire line in July 2018 near Peachland, British Columbia, I had not set out to write a book about wildfires. I was on assignment for the *Toronto Star*'s BC bureau, and on that first fire I made many mistakes. I showed up without alerting the wildfire service. I didn't have any protective equipment beyond a pair of hiking boots. I didn't even know what Nomex was. I was—like most Canadians—a wildfire neophyte. The learning curve has been steep.

The sheer volume of what I didn't know then has filled this book. The sum total of what I still don't know may yet fill another. Since that 2018 fire in BC's Okanagan wine country, I've covered not only ever-worsening wildfires, but the broad and growing impacts of climate change.

For me, 2021 was a watershed year. In May, I interviewed unhoused Vancouverites in the city's scorching Downtown Eastside during a heat dome that killed six hundred people. The sidewalk reached 56 degrees Celsius under a blistering sun. In July, I photographed homeowners fighting the White Rock Lake wildfire after it destroyed wide swaths of ranching communities across south-central BC. In November, I stood waist-deep in Abbotsford's frigid floodwaters as farmers struggled to save thousands of livestock from drowning. All of this destruction in only half a year. When student journalists ask me why I became a climate disaster photojournalist, I tell them I didn't choose to. The disasters just keep happening.

Over the past nine years, I've come to realize two important things: We are not ready for what is coming and, in many cases, is already here. We've refused to heed experts' warnings for more than three decades.

But for me, there is a deeper, more insidious problem at play: We are not ready because Canadians do not want to be confronted by our own disasters. We largely refuse to look at them and, when we do look, we see only the spectacle. We allow our governments to hide what is happening behind roadblocks and platitudes until they decide whether or not to show us. I've been threatened with arrest while doing my job more times than I can count. I've been detained. My colleagues have been arrested.

I don't see how meaningful action to address wildfires or climate change is possible in Canada until we are willing—and able—to really confront what is happening. We fear what we do not understand, and we cannot understand what we cannot see.

Reporting this book was challenging and—at times—extremely frustrating. Many of the reports and government documents I relied upon are the sort that are routinely made public in other countries. I had to obtain them as leaks from sources who risked their careers to share them because—despite being public documents—in Canada, they are often withheld for years. If they are eventually released, it's usually with the important details redacted.

I also relied heavily on the testimony of dozens of firefighters who took similar risks by speaking to me, on the condition that I not identify them. Government policy often prevents them from speaking freely about their experiences or saying anything that could be construed as being critical of their employers. In some cases where noted, those people are identified by a pseudonym.

In many others, the expertise and stories they shared with me are not referenced directly but were an enormous help to my understanding and my approach to the book overall.

I am deeply grateful to these people, first for their willingness to keep going into increasingly dangerous forests on our behalf, and second, for trusting me with their perspectives and stories about what they saw there.

Unless otherwise noted, dialogue quoted in this book is drawn from formal recorded interviews with sources or from contemporaneous notes I made in the field, and often a combination of both. In some cases where noted, quoted dialogue is based on the recollections of people who were present at the time and, where possible, has been independently fact-checked with the original speaker.

I've also noted the original sources for all documents and research I relied upon. In some cases, general descriptions of technical wildfire terms and jargon like fuel type and firefighting strategies are not directly cited. These are based on my own research, including extensive conversations with eminently patient firefighters and scientists who shared their knowledge with me.

It's also important to acknowledge that I am not a firefighter. I cannot pretend I fully understand their world or the pressures and dangers they face. While I have been fortunate to spend more time on fire lines than many Canadian journalists, my experience pales in comparison to the months and years these firefighters spend risking their lives to help protect the rest of us.

This book is for them.

WILD
FIRE

INTRODUCTION

In the summer of 2023, it seemed like nearly all of Canada was on fire. The season had started early, with fires tearing through Alberta's Yellowhead County in May, scorching thousands of square kilometres and forcing thousands of people, mostly ranchers and farmers, from their homes. Then Nova Scotia exploded, catching an entire province off guard. Recent seasons have made wildfires feel routine in Western Canada, but seeing the Upper Tantallon blaze rip through the bedroom communities of Halifax left many veteran firefighters shaken, afraid of what the rest of the summer would bring.

By the time I arrived on the Tsah Creek wildfire, burning roughly five square kilometres between Vanderhoof and Fort St. James in northern British Columbia, it was mid-July and my home province's turn to burn. I'd driven more than ten hours north from Vancouver to get here. After checking in at the incident command post, I pulled off the highway, parked my aging Subaru in a dusty clearing plowed out by bulldozers, and went in search of a wildland firefighting crew working nearby. My Subaru's dashboard told me it was approaching thirty degrees Celsius outside. It was just a few minutes before 3 p.m.—the peak burning window, as firefighters say. Stepping out of my car's air-conditioned comfort felt like opening the door to a furnace.

Like most photojournalists in Western Canada, I'd covered wildfires before but never from this close. Press restrictions mean we're usually stuck at some roadblock many kilometres away, unable to see much beyond towering smoke columns on the horizon. Or we're forced to wait until weeks later to photograph remnants of destruction when evacuation orders finally lift, long after the fires themselves have gone out. Most Canadians have never seen a wildfire up close, and many hope they never will. What we know of them comes mostly from the snippets of cellphone footage shot by fleeing, often panicked civilians. We see them driving through walls of black smoke and swirling embers as stands of trees belch flames high into the sky, the roadside ditches ablaze with what look like the fires of hell.

Now, walking down the road in this particular fire's wake, I was struck by the eerie quiet of the place. Animals had fled. No birds squawked to announce my presence. Around me the blackened forest smouldered. The entire forest canopy was gone; the spruce and pine trees stood naked, their needles burned away. I could see for almost 100 metres in every direction. Up ahead, beyond my line of sight was the fire front itself, marked by the greyish-white smoke boiling out of the treeline. Through the haze around me, a few hundred metres away, I spotted several trucks parked in a tight formation.

At the gravel pit staging area I found a handful of the thousands of firefighters brought in to help Canada battle what was by then already the worst wildfire season on record. These particular firefighters were a crew of smokejumpers from Alaska, elite athletes who parachute onto remote or inaccessible fires with the tools, food and camping gear to support themselves for days at a time. They often sleep near their fires. Resources were already stretched so thin across BC that summer that this crew of smokejumpers had been combined with firefighters from the province's own parattack firefighting unit, based in Fort St. John. Despite their specialized expertise, the combined Canadian/Alaskan crew had been driving to nearby fires instead of flying to remote ones, because the province couldn't afford to hold them in reserve.

A wildland firefighter in a bright red shirt and orange hard hat stood on a rocky ridgeline in front of me, silhouetted against the smoke-choked sky. She waved. I waved back and scrambled up to join her.

"Congrats," she said, in a lilting Australian accent. "You've found the head of the fire." At that moment, the wind gusted and swallowed us in dense smoke. The fire's main column had blown down on top of us. Through the haze, I could make out the dim glow of treetops torching on the other side of the gravel pit.

For a few seconds, the whole world vanished. I couldn't see six metres in front of me, but my new firefighter friend was unfazed. She called in the condition change on her radio then turned to face me. She told me her name was Meg and she was a lookout for the BC Wildfire Service unit crew assigned to the fire. Her job that afternoon was to monitor the head of the fire and radio its behaviour back to the rest of the crew so they'd know where it was, what it was doing and whether a sudden wind gust might be driving it right towards them. As she watched and reported the fire, the rest of the crew worked up its flanks, doing what wildland firefighters do to contain blazes: digging fire guard by hand through the forest floor with Pulaskis (think part axe, part heavy-duty garden hoe), laying out pumps and hoses, spraying down spot fires that had jumped over their lines; and generally trying to steer the flames away from the highway. Wildland firefighting is grinding work. Unlike their structural firefighting counterparts, wildland firefighters typically do not arrive at a blaze with trucks full of water. Instead they have to read the landscape and make due with whatever natural sources they can find, whether that's a thundering river, a stagnant swamp or, sometimes, even puddles.

Meg was from Australia, and her family, like many in that country's fire-prone state of Victoria, had a long history as volunteer firefighters with the Country Fire Authority. She recognized my yellow fire helmet, with its CFA emblem, immediately. To a firefighter, I must have looked like an odd sight: yellow CFA helmet from Australia, red Nomex shirt from BC, and green Nomex pants from the US Forest Service. I started to explain to Meg that I'd spent the previous winter with volunteer bushfire brigades in Australia, who had let me keep the helmet as a souvenir. But before I could finish, a helicopter swinging a bulging water bucket beneath it hammered over the treetops towards two of the nearby smokejumpers, ready to drop. I stammered a quick goodbye and raced over, hoping not to miss what I was sure would be an opportunity for strong photos.

I caught up with the pair just as one of them, smokejumper Tyler Moylan, shouted instructions into his radio. He craned his neck to keep the helicopter in sight as it hovered almost directly above him, barely 30 metres away. Over the roar of the helicopter's engine and the howling downdraft from its rotors, I couldn't hear what he said. Whatever it was must have worked, though, because as soon as he said it the pilot opened the bucket's giant drawstring. It released a towering thousand-litre column of water that crashed down in front of us with so much force that I could feel the whoosh of air fleeing the deluge. It nearly obliterated the small burning spruce tree and soaked the ground around it, right on target. Moylan flashed him a thumbs-up. Through the machine's window, I caught the pilot's brief wave of acknowledgement before he punched the throttle and lumbered into the sky towards a nearby lake to fill up for another run.

Moylan and his colleague Steve Lozano attacked the dirt around the smoking tree with shovels and, to my surprise, the earth responded with a gush of steam and smoke. Though it was just a small spot fire, it had already burrowed underground. One drop from the helicopter had smothered the flames and cooled it down, but fire is persistent. Before this one gave up, Moylan and Lozano would have to dig it out by hand, sometimes up to their elbows, scattering its embers until everything was cold to the touch. Firefighters call it cold-trailing. They would have to do it for hundreds of hectares around the perimeter of this fire alone, and there were hundreds more across the province, and thousands across the country.

Once they finished with the burning spruce, Moylan squared his shoulders, shifting his heavy line pack and scanning the nearby treeline for signs of fresh smoke. Amid the general haze produced by the main fire, which was still burning on the far side of the gravel pit, the new *smokes* were often hard to spot—just a wisp of grey amid the lighter greyish-orange backdrop. They indicated new spot fires, lit when embers from the main fire drifted further ahead. Sometimes a shred of burning bark, needles or even a pinecone lofted into the sky by the main fire's updraft, would carry on the wind before settling back to earth into a clump of dry grass or the waiting lower branches of the tinder-dry trees all around us. Wherever the burning ember landed, another fire would start—small at first, but easily capable of spreading quickly. Spot fires are one of the most common and most dangerous ways a wildfire can spread. Unwary firefighters might not notice the tiny flame that skips over their heads, igniting a fire at their backs while their focus is on the bigger blaze in front of them. On fire lines across the world, firefighters have been burned, and some killed, by spot fires that outflanked them and caught them off guard.

As soon as Moylan or Lozano found a new spot fire, they'd repeat the process of scraping away the dirt, needles, twigs and other debris around it (what firefighters call the *duff layer*) with their shovels and Pulaskis, creating a small rudimentary fire guard to keep the spot fire from spreading. Then they'd get on the radio and call in another water drop, guiding the helicopter pilot in so he could hit his target, often with pinpoint accuracy. Then they'd dig out the remains, extinguishing whatever smouldered in the duff layer of the forest floor, and move on to their next target. I chased spot fires with Moylan and Lozano for the rest of the afternoon, eventually getting the hang of the rhythm as they followed embers across the sky, scanned for the forest for smokes, ran them down, called in more bucket drops. As the winds slowly faded and the smoke began to lift, they retired to the centre of the gravel pit to regroup for a night operation: a small-scale planned ignition. In layperson terms, they were going to set part of the forest on fire, on purpose.

Once upon a time, fire agencies called these operations *controlled burns*, but the name is a bit of a misnomer. Once an intentional fire is lit, it's not always possible to control what it does. All firefighters can do is prepare effectively, trying the best they can to predict what the fire will do given the prevailing winds, fuel moisture codes, overarching drought codes, relative humidity and fuel (the type of trees, shrubs, grasses and other flammable material) they're trying to burn.

A fire needs three things to burn: fuel, oxygen, and a source of ignition. The goal of a planned ignition can vary, but at their core, ignition operations are about removing one of the fire triangle's three sides: its fuel. By burning up stretches of forest under favourable conditions, firefighters can prevent it from burning during unfavourable ones. A successful planned ignition is one of the best tools firefighters have to contain an out-of-control wildfire, but these ignitions are risky, and they can go sideways.

Holding operations are a good backstop, where supporting firefighters are stationed or patrol the edges of a burn perimeter, scanning the green—unburned—side of the fire guard for spot fires and knocking them down quickly: essentially holding (or trying to hold) the fire on the black—the burned—and within predetermined boundaries. But if a fire crew's ignition goes wrong because they misread any one of a dozen different variables even a little, there is likely no stopping the fire. Hence why, in recognition of this lack of true control, fire agency bureaucrats now prefer the drier term *planned ignition*.

Understanding how these tactics work—how to read the land and the forest and the sky, and how to set fire to a swath of forest safely—takes years to learn. It can't be taught in a classroom alone. An experienced firefighter can read the relative humidity by licking their thumb and rubbing it against their index finger for a few seconds. They can look at a stand of trees and judge that *"While the manual says it should be Fuel Type C4* (immature Jack and lodgepole pine), *I think it'll burn more like C7* (ponderosa pine and Douglas fir)." That has direct implications for where firefighters place their crews, what ignition pattern they use, and whether their fire is likely to get away from them or not. It's art as much as it is science.

The plan for that night at Tsah Creek was simple: the firefighters would work north along the logging road using drip torches to ignite a surface fire that would burn up available surface fuels (the flammable needles, dry leaves, twigs, shrubs and branches that litter a forest's floor) along the fire's eastern flank. As the firefighters gathered for a planning briefing, I walked over and introduced myself to one of their squad leaders, a bearded smokejumper named Jake Murie.

Murie is tall, with sloping broad shoulders and the firm handshake of someone who fights fires all summer, then rafts whitewater and climbs mountains all winter. He proceeded to sketch out the nearby terrain, highlighting water features, a small pond and swamp, a ridgeline and their main fire guard—an old logging road that had been improved with the help of heavy equipment, plowing away any shrubs and plants that had grown over, removing everything down to unburnable mineral soil. Assuming all went according to plan, the following night they would burn the western flank through much heavier timber, with the goal of driving the fire towards the gravel pit where it would—hopefully—be boxed in.

As dusk fell, the hodgepodge crew of Canadians and Alaskans got the burn underway. They took turns walking through the forest up to a few dozen metres from the road, dribbling a burning mixture of diesel fuel and gasoline from the pigtailed spouts of their drip torches. They worked in a grid pattern: creating small boxes of flame a foot or so wide, which combined into bigger boxes a few metres wide, interlocking with other boxes-of-boxes to create a quilt of fire across the forest floor. The goal of such a pattern is to have the tiny flames grow towards each other fast enough to build the heat and momentum needed to consume the layers of leaves, dead branches, shrubs and other surface fuels, but not so fast that the forest explodes into a wall of fire.

This, more than anything else, is what I had driven more than a thousand kilometres to see. From a photographer's perspective, wildfires really only look good at night, when their flames light up the forest and sky with an unearthly orange glow. Embers streak across the sky, leaving behind trails of light as the camera's shutter snaps closed. It is simultaneously beautiful and haunting, thrilling and yet strangely calm. Once you've seen it up close, you want to keep seeing it. You have, like me, caught the fire bug.

I MET UP WITH MURIE AND HIS CREW AGAIN THE NEXT NIGHT, AS they prepped for the larger, more challenging burn. Ironically, the biggest hurdle they faced was that relative humidity had risen, and the forest floor fuels—what wildland firefighters call one-hour and ten-hour fuels, because of how long they take to lose or gain roughly half their moisture—were no longer as volatile as they'd been over the previous few days. Murie worried he'd have a hard time getting his ignition going, which could mean losing the precious burn window of calm winds and moderate humidity he needed. If the burn failed, it would leave the western edge of the fire's perimeter less secured, giving the fire an avenue of escape towards the nearby highway.

Because tonight's burn was bigger and carried more risk, the smokejumpers would be backed up by a British Columbia unit crew: twenty firefighters divided into teams of four, positioned along the edges of the burn to do holding operations.

As the smokejumpers and Columbia crew leaders discussed their tactics for the night, a pickup truck rolled into the gravel pit. A local contractor had driven out from town with boxes of pizza for the firefighters. Everyone grabbed a slice and took a short break. Moylan reclined against the passenger door of the contractor's truck when—without warning—a puppy's face burst through the open window and snatched the pizza from Moylan's lips. Dozens of firefighters burst into laughter, taking a moment to snuggle with the adorable pup before hefting their drip torches and heading into the forest.

Later that night, as the ignition was well under way, I followed Moylan and BC Wildfire Service firefighter Charlie Helton as they worked their drip torches through dense thickets of shrubs, over tangles of fallen logs and dead branches. Only a few dozen feet away, stands of pine trees were torching together—fire tearing up along their trunks, leaping branch to branch, climbing the trees like a ladder. The sound a stand of trees makes when it torches is hard to describe: References to jet engines or rumbling trains or a combination of both come close. To me, it sounded like the heaving of a giant's chest, hot breath whooshing in and out as it gathered itself to roar.

At that moment, Murie's voice crackled over our radios, calling us to a halt. The fire's intensity was building faster than he liked, and other burn crews closer to the road were trailing a bit too far behind us. So, we stopped. Moylan found a fallen log and sat down like someone about to eat lunch in a city park. Helton sat across from him, and I suddenly found myself listening as the two discussed the finer points of surfing in Mexico—all while the fire around us continued to shoot flames a hundred feet into the air. After a short break, Murie called back with the green light to continue. Moylan and Helton got to their feet and continued with a nonchalance I found both comforting and also a little unnerving.

Back on the roadway, I found Murie again, pacing along the fire guard, radio in hand. What he read in the flames I couldn't exactly say—all I saw were walls of fire belching embers high into the night sky. Around us other Columbia crew members stood and marvelled at the sight. Murie turned to me with a smile. "Pretty incredible, isn't it?" he asked. "Sure beats having a day job," I replied, raising my camera and carefully framing him against a sheet of flames as another stand of trees went up. The burn continued through the night and went off without a hitch. As I crawled into my sleeping bag back at the fire camp sometime after 2 a.m., I had trouble falling asleep—not from nerves or fear, but from the rush of seeing something so few people get to see. It's no mystery to me why, despite the danger and the stress, wildland firefighters love their jobs so much.

I didn't realize it at the time, but reporting from the Tsah Creek fire has become foundational to my understanding of modern wildfire and of what is changing in our forests. Wildfires are often hard to describe—the immensity of their destructive power tends to be visible only for moments or minutes. A stand of trees torches hundreds of feet into the air; a house is engulfed and reduced to ashes in no time and then—nothing. Blackness and a lingering haze. When it's over, even while surveying the charred remains, it's possible to wonder if what you just saw really did, in fact, happen at all. In these cases, a photo is often the only thing that can really capture a fire's power; permanently freezing sheets of yellow, red and orange flame. Fixing them in place is the only way to study them long enough to really comprehend their fury. This is the wildfire that the public has slowly come to know: the all-consuming, ever-destructive monster hungry for more.

These images are vital to our understanding of the crisis we face. We fear what we don't understand, and we can't understand what we cannot see. But without important context, these images often drive simplistic narratives about wildfire as wholly bad, a thing we need to tame, control, stamp out. The truth is much more complex, and the spectacle hides our own complicity in the destruction now frequently wrought not by traditional wildfires, but by megafires. As overwhelming as it was to a relative rookie like me, for folks like Murie, Moylan, Helton and thousands of others across North America who fight wildfires for a living, the Tsah Creek wildfire was a pretty unremarkable work-a-day fire—the kind that generations of wildland firefighters have perfected the art of containing, suppressing and putting out. In fact, we've collectively gotten so good at it over the past century that we have—or rather had—virtually eliminated wildfire from the landscape. And, as we are slowly learning, that is not a good thing: fewer routine healthy fires have paved the way for the destructive monsters that are essentially unstoppable.

But fire doesn't just destroy, it restores. Many ecosystems across western North America are what's called *fire-adapted*. It means they've evolved not just to coexist with fire, but to rely on it. The seed cones of Jack and lodgepole pines only open when exposed to the heat of a wildfire, making it a necessary part of the forest's life cycle. Fire thins out forests, creating gaps in otherwise cloistering canopies, letting in sunlight so all manner of understory grasses and shrubs and berries can thrive. These in turn become food for animals, like bears and deer, and even people. Photographs from 110 years ago, at the founding of what became Jasper National Park, show the Athabasca River Valley not as the dense, tree-choked forest it is today, but as a mosaic, with open meadows of grass dotted by stands of trees all varying in age. The photographs are old and grainy, but even still you'd be forgiven for thinking you were looking at a golf course, not a Rocky Mountain forest.

Indigenous nations in North America learned how to coexist with fire thousands of years before the first settlers arrived. Indigenous peoples' use of strategic, cultural burns was widespread and accomplished many important tasks, including improved berry crops, the creation of easier hunting grounds and community protection. European colonizers outlawed those practices and built a whole industry around trying to put out wildfires at almost any cost.

But wildfires are as inevitable as rain, and just as necessary for the health of many North American ecosystems. Trying to eliminate them is as futile as trying to dam the heavens, and such efforts have helped set the stage for a new kind of wildfire we are not prepared to face. Confident in our ability to put out all wildfires, we grew complacent. Many of us put wildfires out of our minds and put off doing the important, simple tasks that make homes and towns more resilient, safe in the belief that if a fire ever did threaten, an army of Nomex-clad twenty-somethings would show up to save the day. We blame them when they can't, not realizing that what we're asking of firefighters today is far more difficult—and more dangerous—than what we expected of their parents and grandparents. The rules of the game have changed—because we changed them. Wildfire behaviour experts call it fire debt—the longer a fire dependent landscape has gone without fire, the deeper the debt. Obsessively extinguishing healthy fires doesn't actually stop them from happening; it just adds more forest to the unsettled tab and worsens the outcomes when that debt finally comes due.

Two years after the Tsah Creek fire, I would see the devastation wrought by this new kind of monster. In Jasper National Park, it wasn't the blocks of destroyed homes or the blackened husks of burned-out cars that got to me. As painful as that destruction is, I'd seen it before. It's become a near-constant presence in our summers now. What most shocked me was seeing the mountainsides just below the famous Marmot Basin ski area, only a few precious kilometres from the town itself, where a suspected fire tornado touched down. Across an area nearly fifteen kilometres squared, not a single tree was left standing. Most had been ripped from the ground, roots and all, and scattered like straw in wide, curving arcs. Fire-generated winds had topped 180 kilometres per hour, fast and hot enough to flay the bark off trees, leaving them gleaming white, dead, and so disfigured forestry experts struggled to determine their species.

Sometimes a wildfire can burn so hot it can sterilize the soil, killing the roots and mycelium networks, the seeds and microorganisms a forest relies on to recover. Below Marmot Basin, the soil isn't just sterilized; it's completely gone, burned down to the bedrock and boulders, many of which shattered in the heat. It doesn't look like the aftermath of wildfire. It looks like the site of a nuclear blast.

What Tsah Creek and Jasper and a dozen fires in between taught me is that there are wildfires, and then there are *wildfires*. Our blanket fear of the former, and our obsession with eliminating them, drove us to create the conditions for the latter, literally stoking our forests with fuel. Human-caused climate change is dumping gasoline on the pile and ramping up the lightning that so often sets a fire ablaze. And yet we still talk of wildfires like they are all the same, all equally bad, and all equally possible to put out if only we had enough water bombers and firefighters willing to risk their lives. My time on the front lines gave me a glimpse into our collective future and upended what I thought I knew about modern wildfires. I set out determined to understand our fastest-growing blazes, our narrowest close calls, and the monster fires that really should scare us. This book is an account of that journey and—hopefully—an argument for change. We can either rethink most of what we know about wildfires, and fast, or we can stand by and watch as our communities continue to burn.

CHAPTER 1

K21620

It started small. A lightning crack and a hesitant wisp of smoke on a remote BC mountainside. Though it would undergo many transformations before its end in the winter mountain snows, when wildfire K21620 was first detected on July 12, 2023, it was barely 0.2 hectares—smaller than most playgrounds.

All fires, when they first start, are fragile things. It doesn't take much to snuff them out: a drop in wind or temperature, a bit of rain, too much humidity. A young wildfire faces extreme odds. Most struggle just to be born, even in a country as heavily forested as Canada.

Like K21620, about half of all wildfires in Canada are started by lightning. The other half—escaped campfires, downed powerlines, sparks from a dirt bike's exhaust or even a backhoe blade striking a rock—are rolled into the broad category of "human caused." Under normal conditions, these fires are typically easier for firefighters to handle. A careless cigarette butt flicked into tinder-dry grass can easily spark a wildfire, but its proximity to humans is often its death sentence. Such fires are relatively easily detected, and often quickly contained or snuffed out.

But K21620 had one distinct advantage: Lightning fires are, by their very nature, harder to detect. They can stay hidden in vast Canadian forests far from human eyes, smouldering for hours, days or sometimes weeks before the fire triangle aligns to breathe life into the embers. Across Canada there are over two million lightning strikes every year, the vast majority of them in June, July and August. Of those only a small fraction—an average of around four thousand, or less than a quarter of a per cent—actually spark a fire. Once lit, these tender flames need careful nourishing: the right balance of heat, humidity, wind and fuel.

As statistically unlikely as any individual lightning fire's survival may be, the potential damage they can do is rapidly increasing. In 2023, there were around seven thousand total wildfires, about 25% fewer than Canada typically sees in an average year. Of those, just over half were caused by lightning, but they were responsible for more than 90% of the staggering 172,036 square kilometres that burned, an area nearly twice the size of the island of Newfoundland. The impact of lightning fires is increasing while that of human-caused fires continues to decline year-over-year. We're getting better at not starting fires in the first place, and when we do accidentally light them, they are the easiest to quickly put out.

As climate change continues to warp our weather patterns, scientists predict several things will happen. First, we will see more frequent droughts like the multiyear one that left Canada parched in 2023. Rainfall will become more erratic. We will see longer periods with no rain at all, followed by more intense rainstorms that, while providing a temporary reprieve, are less likely to give the kind of deep-soaking drought relief needed to shift the wildfire risks.

At the same time, a warmer climate means more thunderstorms, particularly at the height of summer's wildfire risks. According to a 2014 report in the journal *Science*, North America can expect to see a significant jump in lightning strikes per year for every 1 degree Celsius of warming above pre-industrial levels. That means a potential 50% increase in lightning strikes by the end of the century. During the devastating 2023 wildfire season, average May–October temperature across Canada was 2.2 degrees Celsius higher than normal. If two million lightning strikes amid record-breaking drought caused 3,500 fires, imagine what three million strikes into even drier forests could do.

All of this will result in more wildfires, a scenario that will require massive investments in our firefighting capabilities in order to manage. If that were all we were facing, simply spending billions more on wildfire suppression might be enough. After all, North American wildfire services have essentially perfected the art of suppressing the kind of run-of-the-mill wildfires that our forests evolved to rely on. The common political calls for ever more firefighters, more water bombers, essentially more *firefighting* would be a viable solution if the problem was just *more* wildfires. But this is not the case.

Regular wildfires during historically normal weather are not what's responsible for the worst destruction we see. Mega destruction is the province of truly monster wildfires, ones that need a particular Goldilocks environment to develop. For a wildfire to really take off, it needs what firefighters call *fire weather*. This term describes the coincidence of factors that create the most dangerous fire conditions—temperature, humidity, wind speed and recent precipitation.

The worst fire weather happens on what are called *crossover* days. This is when the relative humidity dips below the ambient air temperature. To gauge the conditions under which explosive fire behaviour should be expected, firefighters also often refer to the 30-30-30 rule: relative humidity below 30%; ambient air temperature above 30 degrees Celsius; and expected wind speeds above 30 kilometres per hour. The more extreme the crossover, the more severe the potential fire behaviour. The most destructive fires in Canada occur during a relatively few extreme fire weather days. It's what takes a regular work-a-day wildfire and turns it into a potentially town-erasing monster.

Without extreme fire weather, most wildfires in Canada can be handled effectively. They tend to grow slowly enough that firefighters and emergency officials have time to respond, declare evacuation orders when necessary and muster firefighting techniques honed over more than a century that are—generally speaking—enough to keep them from destroying homes or other infrastructure.

But as with lightning strikes, a warmer climate means more fire weather days. Between 1980 and 2010, most of Canada saw an average between zero and five fire weather days per year. Northwestern Alberta, Northeastern BC and the southern reaches of the Yukon and Northwest Territories—long the most fire-prone landscapes in Canada—saw an average of ten to fifteen per year. Research published in August 2024 by Natural Resources Canada shows in a best-case scenario, the annual frequency of fire spread days is modelled to increase by 35% by 2050. The worst-case scenario is an increase of 400%.

Think about it this way: The now-infamous Fort McMurray wildfire exploded on May 3, 2016, and did the majority of its damage to the city within sixteen hours, burning more than 1,600 homes by the following morning. The McDougall Creek wildfire that destroyed more than 200 homes in Kelowna and West Kelowna in 2023 did so in just thirty-six hours. The 2024 Jasper wildfire went from ignition to levelling a third of the town in less than two days. The total destruction wrought by these fires, nearly three thousand homes and businesses in all, happened over a combined three and a half days of extreme fire weather, the current national average. Imagine what ten days per year of this might look like. Imagine more than a month of them every summer.

More than the total area burned or the sheer number of blazes, what is really changing in our forests is the potential severity of the fires themselves, infernos that can move with a speed and ferocity often unprecedented in modern wildland firefighting. It's blazes refusing to quiet down at night like they reliably used to. It's week after week of crews putting in back-breaking work digging fire guards, running hose lays, conducting backburns, only to have one fickle day of fire weather erase all their efforts.

These changes are happening so fast that even institutional knowledge struggles to keep up. As one frustrated crew leader told me in the midst of the 2023 season, "if you haven't fought fire on the ground since 2017, you don't know what we're dealing with. We're dealing with hell."

K21620, AS IT WAS FIRST CALLED, STARTED ON A REMOTE MOUNtainside above the eastern shore of Adams Lake, a long, deep body of glistening cold water in the Monashee Mountains of BC's southern interior, about an hour's drive east of Kamloops. The lake—the second deepest in the province—is known for its fishing, water skiing, windsurfing and even some scuba diving. It's home to a small collection of communities clustered at its southern end, mostly cottages and cabins, including the communities of Woolford Point and Dorian Bay, where the Adams River flows down a set of rapids popular with whitewater rafters and spawning sockeye salmon and into Shuswap Lake.

For many fledgling wildfires, having won the lightning lottery, managed to spark and struggled to survive long enough for a breath of dry wind to fan their flames, they face their most formidable foe yet: a wildfire initial attack crew. Borne across the landscape in helicopters, or pounding over forest service roads in pickup trucks, initial attack crews (IA, for short) are the shock troops of wildland firefighting, and they are brutally efficient. They are small (typically only four or five people), light and agile. They often arrive at a fire with enough supplies to operate independently for days in the forest before backup arrives.

Once on the ground, they will size up a fire and set to work digging a hand guard; scraping a foot-wide strip of forest floor around the fire down to mineral soil, removing anything flammable like twigs, roots or blades of grass. A clean hand guard is a wildfire crew member's pride; it's considered bad manners to tread directly on it. Crews also evaluate the landscape for water sources and plan the layout of pumps and hoses if they need them, all the while scanning for dangers like burning debris rolling downhill, pockets of extremely flammable vegetation and dangerous trees that can fall with lethal force.

Initial attack also often includes water bombers or helicopters, or both. In the early stages of a fire, these can help cool and slow a fire long enough for IA crews to get them surrounded with guard and start the laborious process of extinguishing them by hand. But like Moylan and Lozano on the Tsah Creek fire, it's not enough to just dump a bunch of water and call it a day. Even for small fires to be declared extinguished, firefighters must dig out smouldering roots and debris and soak everything until it's cold to the touch. Water bombers alone cannot put out wildfires.

British Columbia's IA crews are among the best in the world, eliminating nearly 90% of the baby fires they fight before they reach one hectare (or 1 ha, pronounced *haw* in firefighter parlance) in size. Even amid the 2023 wildfire season, which shattered records across the country and killed six of their colleagues, the BC Wildfire Service's initial attack success rate "fell" to 88%—a straight-A grade in any other discipline but a decline that still left many proud veterans smarting.

K21620 did not face an initial attack crew for most of its early life, nor did it see any significant fire weather at first. That would come weeks later. Instead, it was triaged as a lower priority fire, one that officials assumed could be safely monitored and left largely to its own devices. Helicopters were dispatched to do what they could from the air but had little practical effect. By early July, the BC Wildfire Service was stretched beyond belief by an unrelenting season. Many firefighting crews were already on their fifth or sixth deployment of the summer, with many more weeks and hundreds of other fires still to fight. Against that backdrop, K21620 seemed virtually benign. Beyond that, access to the rocky, cliff-strewn mountainsides where K21620 was burning was a challenge, and for two weeks the fire service believed it posed little risk to anyone.

I first saw it on July 27, when I drove up a logging road on the far side of the lake, found a decent vantage point, and spent the evening making long-exposure photographs as it burned gently on the mountainside across from me, the wind blowing it harmlessly north and away from lakeshore homes. It was a peaceful evening. After about an hour, I packed up my gear and headed back to my hotel. *This fire's not going anywhere*, I remember thinking to myself.

CANADA IS A COUNTRY ALMOST PERFECTLY DESIGNED TO BURN. Many Canadians think mostly of either "the city" or "the forest," but the reality is much more complex. There are eight different forest types in Canada, organized into twenty ecozones; sixteen different wildfire fuel classification types (the specific makeup and density of trees, shrubs, leaf litter, dead branches—everything in a forest that can burn); and a practically infinite number of potential combinations of fuel, weather, and topography. The largest of these forest types, the boreal forest, makes up roughly 70% of our trees, and it blankets more than a third of the entire country in a mix of black spruce, Jack pines, and other highly flammable trees. With 5.5 million square kilometres, Canada accounts for more than 25% of the world's boreal forest, an ecosystem that relies on wildfire to regenerate.

And it's not just that this wild land exists *out there* in the backcountry somewhere beyond our manicured lawns and decorative cedar hedges. Canadians increasingly live in these forests. Urban sprawl, the growth of Ontario's cottage country and other forms of development are continually expanding what experts call the *wildland-urban interface* or WUI (unfortunately pronounced *WOO-ee*), the places where human development meets the forest. Nearly five million of us—about 12% of Canadians—already live in the WUI. We're not just talking about rustic cabins in the forest either. Whole neighbourhoods of North Vancouver, Thunder Bay, the Muskokas, Halifax and more sit in the WUI zone, surrounded by and run through with trees. Across the country there are more than 324,000 square kilometres of homes, businesses, playgrounds and hockey rinks potentially at risk from wildfires. If current global trends hold, that could grow by another 25% in the next two decades.

We are building deeper and deeper into the forest just as our forests are becoming far more flammable. The seven worst wildfire seasons in Canadian history have all occurred in the last decade. In the 1970s, Canada saw an average of 8,900 wildfires per year, which burned an average of twelve thousand square kilometres—an area larger than Cape Breton Island. Since then, the number of fires per year has declined slightly in Canada, but the total area burned has nearly tripled to a ten-year average of twenty-nine thousand square kilometres. We are having fewer fires, but they are burning far more area per year. If you count only 2020 to 2025 (which includes four of the ten worst seasons in recorded history) the average rises to a staggering sixty thousand square kilometres per year, five times more than the 1970s. That's an area twice the size of Vancouver Island burning *every summer*.

The cottages and towns of BC's North Shuswap region, including those along Adams Lake, are classic wildland urban interface zones. Houses and cabins sit perched on hillsides overlooking seemingly pristine lakes, shrouded from their neighbours by the embrace of pine and cedar boughs leaning, in some cases, close in over their decks. The houses along the lake seem to cling to the shoreline, built long before concerns about riparian areas and fish habitat led to legislated set-backs from the lake. The charm of these mountain getaways nestled into the forest is what keeps families and tourists coming back year after year. It's also what makes them dangerously susceptible to wildfires.

By late July, K21620 had been dubbed the Lower East Adams Lake wildfire, and it had many tactical advantages. It sparked on the side of a remote mountain in the midst of a historically bad drought and went undetected long enough to quietly establish itself. It also had the high ground, burning in steep, rocky terrain shot through with cliff faces, making it too dangerous for ground crews to attack directly. Water skimmer planes (the smaller, more agile cousins of Canada's famous yellow water bombers) were dispatched to do what they could from the air, with little effect. The fire had established a beachhead and successfully defended itself against aerial attack, and people living nearby began to worry.

The Adams Lake fire had become part of a complex of three fires sharing its name, which included its twin, the Bush Creek wildfire, which was burning in similarly challenging terrain on the far side of Adams Lake, and the Rossmoore Lake wildfire, burning sixty kilometres away to the west in a mix of heavy Douglas fir and ranchland about ten kilometres south of Kamloops, BC. All three were started when lightning storms rolled across the region. All three fires that took days to discover, catalogue and initially respond to. By late July, the fires were large and complex enough to require their own specialized management teams, but even with thousands of additional firefighters flown in from Australia, Brazil, Costa Rica and Mexico to help, there were not enough resources available. Instead, every incident management team (IMT) that got assigned to the Adams Lake complex would have to do its best to fight all three at once for the duration of their two-week deployments.

First up in rotation was Incident Management Team 5, one of the BC Wildfire Service's most experienced IMTs. It was helmed by Incident Commander Mark Healey, a veteran firefighter with a brusque demeanour and a carefully waxed handlebar moustache. He has a reputation for no-nonsense efficiency and a willingness to take risks calculated against the depth of his thirty-plus years of experience. When they take command of a fire, IMT members don't do the front-line firefighting themselves. Incident commanders like Mark Healey spend most of their time not on the fire line itself, but inside ATCO trailers full of maps, logistics reports, radios and computer screens. They work like a plug-and-play command headquarters, collecting the heads of operations, logistics, planning and even payroll sections into a tight-knit team that works together, often for years. They have their own dedicated operations chiefs, fire weather behaviour analysts and communications officers. When IMT 5 took over the Adam's Lake complex, Healey inherited responsibility for deploying fire crews, helicopters and heavy equipment the way a battlefield general deploys their infantry, artillery and armoured divisions, and air force.

Fighting even one wildfire is like spinning multiple plates in the air at once, constantly evaluating and juggling risks: both the ones you can see, and the ones you can't. The only way to make wildland firefighting 100% safe is to not do it at all. The best any firefighter, crew leader or incident commander can do is try to stay aware of everything that's going on and to mitigate the risks to their crews as best they can.

Healey wouldn't be fighting one battle; he would be fighting three.

In the days after the Adams Lake, Bush Creek and Rossmoore Lake fires ignited, it was Rossmoore that appeared to pose the biggest risk. In mid-July it exploded, launching a towering smoke column into the sky that rolled towards Kamloops like a thundercloud, blocking out the sun. Ranches and homes in the nearby Knutsford community were put on evacuation order. At its height, the Rossmoore fire had hundreds of firefighters assigned, including crews from Mexico, Costa Rica, Brazil, Australia, the US Forest Service and two Canadian army helicopters. Fire crews in BC that summer had already seen other blazes make ten- and twelve-kilometre runs, easily covering the distance to Kamloops itself and its one hundred thousand residents. For weeks, Rossmoore Lake was the focus of Healey's incident command, tying up hundreds of firefighters, heavy equipment, helicopters and even local volunteer wildfire brigades, who were getting their first taste of real campaign firefighting.

Compared to that, the Adams Lake fire appeared to pose very little risk at all, burning away gently on a rugged mountainside as it had for two weeks, with winds blowing it north away from homes and people. Even so, the BC Wildfire Service was keeping an eye on it and deployed helicopters on what one veteran firefighter called "political bucketing" missions—conspicuously flying in and out of residents' waterfront views day after day. The pilots would scoop up 2,500-litre buckets of lake water from between homeowners' docks, waterslides and anchored motorboats, and drop it a few short kilometres away on the now weeks-old wildfire, demonstrating that the fire service had not forgotten about them.

By late July, Canada had been in the grip of its worst ever wildfire season for more than three months, but that didn't change the frustration of local residents in the North Shuswap and along Adams Lake, who watched as the fire above their lakefront homes slowly grew as the weeks rolled by. One helicopter doing what amounted to PR laps wasn't enough to impress the anxious residents of Adams Lake. They felt the fire service needed to do more and continually said so. They spent days sending messages to the local regional district office and to the BC Wildfire Service demanding more action on the fire.

On July 20, homes along Adams Lake were put on evacuation alert, and more aerial firefighting was promised. The neighbourhoods of Dorian Bay and Woolford Point are isolated, a collection of mostly cottages and other modest properties strung along the lakeshore with steep, rugged slopes rising above them, and accessible only by boat or by ferry from across the lake. If an evacuation was ordered, it could take close to twelve hours to ferry everyone to safety.

Pushed by prevailing winds, the fire continued burning uphill and away from homes and the lakeshore during the day, occasionally creeping back downhill at night on gentler winds caused by the dropping mountain air temperature. Firefighters call it *backing down* a slope. The fire was now roughly 3.5 kilometres away from homes and slowly creeping about one hundred metres closer per day.

Residents tried to warn the wildfire service that these favourable southerly winds would change. Local knowledge—something that centralized wildfire agencies often struggle to access and integrate—suggested that typical winds along Adams Lake blow north-to-south, picking up speed throughout the day, not south-to-north like they had been since the fire started. Meeting notes written by local community members describe repeated attempts to convince the fire service of this reality, to little avail.

Frustrated, a group of local residents from the Neighbourhood Emergency Program started calling the BC Wildfire Service directly, demanding more be done to address the fire that was inching closer and closer to their homes.

Across the lake, some ground crews continued fighting the Bush Creek East fire which was easier to reach thanks to a robust network of logging roads that allowed safer access directly to the fire. The fire service also had several heavy-lift helicopters pouring water on the Bush Creek fire, including a massive twin-rotor Chinook, which did laps between a small lake and the fire line, dropping huge columns of frothy white water onto hotspots, trailing rainbows in its wake in the setting mid-summer sun. But the main focus for firefighters was still the Rossmoore Lake fire menacing Kamloops.

That same week, angry residents from the Shuswap, including some from Dorian Bay and Woolford Point, sent what they described as an official "Letter of Concern" to the BC Wildfire Service and the local regional district office, complaining that more had not been done to protect their homes. After days of unanswered phone calls, the BC Wildfire Service held a meeting with frustrated residents on July 26, fourteen days after the fire was first discovered. By this point, trust was beginning to break down. At the community meeting, residents demanded to know why they'd seen so little firefighting activity on the Adams Lake fire, perhaps not realizing the degree to which the fire service was already overwhelmed province-wide.

Two days after the meeting, the fire service ordered heavy-duty sprinkler protection units for the Dorian Bay and Woolford Point neighbourhoods. These have massive diesel-powered pumps that sit on the backs of all-terrain-vehicles and can be driven directly to a water source. They feed far more volume and pressure than traditional wildland fire pumps, but they require more time to deploy. Structure protection specialists spent days setting them up and by July 30 finally had everything in place for a successful test of the system. For a few minutes, water rained down in wide arcs over homes across Dorian Bay, shrouding the neighbourhood in interlocking umbrellas of moisture.

But in wildfires, the situation can change with the winds. A day after the first successful sprinkler test at Dorian Bay, the systems were taken down and sent to another community. Frustrated with the BC Wildfire Service's response, many local residents started taking matters into their own hands, setting up pumps and sprinkler systems on their homes. One person had access to a large water bladder to act as a mini reservoir and set it up in the middle of the road between the lakeshore and their house.

At another joint meeting, tensions began to flare. Angry residents began threatening to defy any evacuation order that might be issued, pledging to stay and fight the fire themselves because they no longer trusted the fire service to protect their homes for them. If they were ordered to flee, getting everyone out on the small lake ferry would be a slow and logistically cumbersome process, but residents pointed to their own motorboats and other watercraft, suggesting they could use those to escape across the lake at the last possible moment if things went sideways. According to local residents, regional district officials responded by threatening to set up a log boom on the lake to cut off that possible escape route and prevent people from returning. Days after that rancorous meeting, the fire would be in residents' backyards.

THE BC WILDFIRE SERVICE'S FIRE BEHAVIOUR MODELLING PRE-dicted the Adams Lake fire would continue burning towards the north and slightly east, away from homes and with only minimal downslope growth at night, as cooler air sank down the mountainsides. Firefighters call it *backing down a slope*, and they expected the Adams Lake fire would pose relatively little risk until at least August 10. Healey deployed his resources accordingly, keeping the majority of his firefighters assigned to the Rossmoore Lake fire.

But predicting fire behaviour is not an exact science. It's a mix of data modelling, weather forecasting, gut-trusting and experience. For more than two decades, fire behaviour modelling in Canada has relied primarily on a software program called Prometheus. It takes landscape data like forest and fuel types; marries it with topography, weather forecasts and the fire danger indicators like the drought code, fine fuel moisture code, and initial spread index; and produces an estimate of how fast a fire will spread, how hot it will burn and in which directions. It's a critical tool for incident commanders like Mark Healey when deciding where to deploy their resources and determining which fires pose the biggest threats. But Prometheus isn't perfect, and like the Greek titan who stole fire from Zeus and gave it to humans, it can't foresee every potential outcome. It's turn-of-the-century technology that relies on broad forest fuel categories and other data inputs that are, in some cases, decades old and out of date. Its most recent update, released in 2022, is considered its "end-of-life" phase. When fire behaviour analysts use Prometheus today, they often have to make gut-check calls, swapping in different datasets their experience tells them may produce more accurate predictions. Used according to its manual, it often fails to predict the kind of growth speeds that modern wildfires are capable of. It failed to predict what the Adams Lake fire would do on August 2, and not for the last time.

As the heat of that day climbed into the thirties, the winds defied the weather forecast. They shifted, swinging around from the south and blowing instead from the north—just as local residents had predicted they eventually would. As the winds pushed the fire back towards the community, it raced down into a creek canyon and up the other side, pushing hard towards the lakeshore homes.

By 2 p.m., winds were topping fifty kilometres an hour, and the fire had blown up, sending a huge smoke plume into the sky that towered over the communities of Dorian Bay and Woolford Point. Residents watching as the fire gathered strength worried and wondered why there appeared to be no sign of firefighters. What they couldn't see was Healey's team racing to redeploy its firefighters from Rossmoore and scrambling to find air attack resources to send when it seemed the entire province was on fire. Publicly, the message from officials was one of calm: despite the intimidating column, the fire was still not an immediate risk to homes.

By this point, the fire had swept along the mountainside above the lakeshore homes and was burning fast towards them. More hours passed as anxious residents took photos from their doorsteps of flames leaping across the hillside above them and of large trees candling. But the community was still only on evacuation alert. Community leaders were encouraging people to flee on their own and many did, despite no formal call to evacuate.

By 6 p.m., local residents in Dorian Bay estimated the fire was less than one hundred metres from their homes, and the few remaining residents geared up to battle it directly with their own water pumps and hoses. Sixteen minutes later, the regional district finally issued an evacuation order, but most residents who remained ignored it and continued trying to fight the flames.

As the fire grew, the wildfire service sent helicopters and a small squadron of single-engine water skimmers that could hit the fire repeatedly for hours. The first of those planes arrived at 6:42 p.m., by which point fire had already surrounded homes in Dorian Bay. The fire service also started pulling fire crews from the Rossmoore and Bush Creek fires, dispatching them to Adams Lake, but getting them mobilized and into action took time that angry homeowners didn't think they had.

As evening fell, the first professional firefighters began to arrive. They went to work with Pulaskis, water hoses and drip torches. The combined force included BC Wildfire Service crews, US Forest Service fire engines and structural firefighters from the Adams Lake fire department. Together, they worked late into the night cutting fire guard, hosing down burning trees and stumps. Silhouetted against the glowing forest, others trailed drip torches through the grasses and shrubs around the bases of trees, sending showers of sparks back into the fire. Luckily, as evening matured into night, the winds also abated, swung around again, and started pushing the fire back up into the hills.

By morning, as the smoke cleared and the sun rose, it became apparent just how lucky Dorian Bay and Woolford Point had been. Scorched trees encircled the area, rising along the ridgeline above their houses and stretching 2.5 kilometres along the lakeshore. Residents who had refused to evacuate continued trying to knock down hot spots with their own equipment, pumps and hoses.

When the day shift crews arrived to take over, they found the line had mostly—thankfully—held. There was some slop over in a few places, but overall their colleagues had succeeded in holding back the advance and bought firefighters a temporary reprieve. Now the day shift had only hours to solidify those gains, reinforcing the fire guards before the fickle winds returned. The fast-paced work continued through August 3, and still the fire refused to back down.

IT WAS BRUTALLY HOT ON AUGUST 4 AS CREW LEADER AWS AL-Mubarak gathered the firefighters of the Fraser unit crew for their daily safety briefing. As a leader, Al-Mubarak carries himself with quiet confidence. While some crew leaders are quick to flash frustration at new firefighters making rookie mistakes, Al-Mubarak uses patience. He prefers conversations over commands, shouting orders only when urgency demands it.

Fraser crew stood in a semicircle amid the trucks, which had been crammed aboard the small ferry travelling to Dorian Bay and Woolford Point. Because of the evacuation order, the ferry was closed to the public; only firefighters were allowed to cross. Because I had already spent days with them on the Rossmoore fire, Al-Mubarak invited me to cross with them, so I parked my Subaru precariously on the loading ramp near the back of the vessel as the ferry set off. The lack of wind had helped hold the blaze in place overnight, but it also meant a suffocating weight hung over the valley, like the smoke itself was trying to smother us.

Most of the twenty-odd firefighters of Fraser crew were on their fifth or sixth deployment of the season, and cracks were starting to show. This wasn't the adrenaline-fuelled adventure portrayed in the BC Wildfire Service recruitment videos. They had already done weeks of grunt work in brutal conditions and exhaustion had taken its toll. Their task on August 4 would be to pick up where the previous night crews had left off. After they landed, Al-Mubarak and a half-dozen members of the Fraser crew went to work with chainsaws, Pulaskis and other hand tools, prepping for another backburn that they hoped would hold the fire at bay.

The trail Al-Mubarak's crew were trying to reinforce ran above a grassy meadow between a string of homes set against the base of the cliff. Before they could burn, they needed to clear a line of guard between the homes and the dense, overgrown brush they planned to backfire. That meant working chainsaws along the black side of the trail, cutting through dense shrubs and thickets of blackberry bushes covered in needles. It was slow, frustrating work. As they unloaded their trucks and got to work, local residents started to gather nearby, eyeing the firefighters from their driveways and yards, unsure what to do.

Standing by the lakeside, Dorian Bay homeowner Ron Hamilton watched as a helicopter rattled overhead. He could see smoke roiling out of the treeline a hundred metres away and flames licking at the edges of his neighbours' field. Having watched the fire creep closer for weeks, Hamilton told me he worried that the fire service wasn't taking it seriously. Like many, he'd chosen to defy the evacuation order, risking up to $10,000 in fines and even jail time, because he wanted to protect his property.

"We felt really nervous," Hamilton told me, shouting to be heard over the clatter of the fire pump he'd set up on his beach. "We really felt that we needed to do something to help or take matters into our own hands in terms of dealing with our own properties." He wasn't alone. Just up the hill from where we stood, another group of homeowners was gathered by the roadside talking among themselves. Anxiety hung in the air as homeowners and firefighters traded wary glances with each other until Al-Mubarak strode over and cut the tension with a handshake. He asked the residents how the fire had behaved overnight, and what they'd already done to try and fight it. It was technically illegal for nonfirefighters to even be here, but given the state of the fire and the lack of resources available, Al-Mubarak wasn't about to rebuke them.

Instead, he recruited them and helped coordinate their efforts alongside those of his professional crew. One of the residents had access to a small backhoe, which could plow a fire break through the shrubs and brambles at nearly twice the speed of a crew working by hand. Al-Mubarak gave him the go-ahead and started rallying his firefighters to follow behind with chainsaws and Pulaskis to clean up the guard.

As he patrolled the line to monitor his crews' work, Al-Mubarak struggled to keep from getting sucked into the work himself.

Being a leader meant staying above the fray, and while it's tempting to pick up a saw or start helping a crew mate who's battling an obstinate blackberry bush, getting drawn in meant potentially losing sight of the larger overall picture and the crew's safety.

Further north along the line, other firefighters were busy with hoses, soaking hot spots in places where the previous night's ignition operations had burned over their lines. The fire had burned right up to the edges of the trail. As Al-Mubarak patrolled past firefighters, offering the occasional pointer or word of encouragement, he heard a sudden cracking sound. He spun around in time to see a small fir tree, not more than twelve centimetres in diameter, slowly topple towards the trail. It snapped and raked its way through its neighbours' branches almost in slow motion before coming to rest across the trail at an angle, hung up in the canopy of other trees across the trail.

Al-Mubarak walked over, reached up with his Pulaski, and hauled the fallen tree all the way to the ground, where it landed with an ominous thunk—surprising for a tree so small and seemingly benign. At nearly ten metres tall, the weight of such a tree is deceiving. It could easily top thirty-six kilograms: more than enough to cause a serious injury if it crashed onto an unsuspecting firefighter or—more likely at that moment—me, an unwary photographer with his face in his camera's viewfinder.

As Al-Mubarak turned to continue his patrol, a young firefighter named Michael came running up to him, worried about another danger tree. This one—a giant Douglas fir—leaned out from the cliffside, fire eating into its base and root system as the blaze worked its way downhill towards Fraser's position. It hung like a sword of Damocles, high enough up the cliff to be out of everyone's immediate line of sight. Right below it, nestled into a crease in the cliff face, was a Mark 3 wildfire pump drawing water from a small creek: Fraser's only reliable water source feeding the hoses that were keeping the rest of the fire at bay.

The pump site would have to be moved, but that meant dismantling the pump and hoses immediately below the burning fir that could fail at any moment. It would be a high-stakes job.

Rather than risk his young crew members, Al-Mubarak decided to move the pump himself. He tapped Michael to help, and they set to work rolling up hoses and moving jerry cans, getting rid of trip hazards outside the danger zone so they could move fast if they needed to. When the moment came to duck in under the tree and retrieve the pump, Al-Mubarak paused for a moment, staring up at the threatening tree. Smoke boiled from the forest floor around them as flames licked at the bases of their trunks.

"If I hear any cracking at all, you run. And Michael? If you run, you run that way," he said, pointing east along the cliff face out of the danger zone. I followed his gaze, spinning plates in my own head, juggling the need to photograph the scene while simultaneously planning out where I'd place my own feet if I had to bolt.

Two hours later, yet another danger loomed. The crew was cutting line along the base of the cliff, preparing for their burn. As the crew and their newly deputized equipment operator pushed eastward, they found a small garage almost entirely obscured by overgrowth. It wasn't on any municipal maps they had, and inside was all manner of junk. An old truck. Most of a broken-down snowmobile. Old TVs. All of it would have to be moved before they could light any sort of backburn.

As the crew continued cutting back the forest around the hidden garage and a nearby home, one of them made a startling discovery. Hidden under dense brambles behind the garage was a stockpile of barrels loaded with what appeared to be aviation fuel. Why it was there or how long it had been was anyone's guess, but for today it meant one thing: the plans for the backburn were off. The crew couldn't risk that cache exploding, even if they did manage to move all of it far enough outside the burn area itself.

All it would take was an errant ember in the tinder-dry grass for it to catch, and they'd already seen what one unexpected wind shift could do.

The crew was frustrated, to say the least. Everything they'd spent the long, laborious day working towards was scrubbed. Some started questioning the wisdom of their orders when Al-Mubarak stepped in and sat everyone down. "Every single person here is exhausted beyond belief," he told them. "All we can do at this point is try to protect these people's homes."

Even if they couldn't carry out the backburn as they'd hoped, they could still try to remove as much fuel as possible along the fire line. "It'll still burn, but let's try to stop it from candling on the edge," Al-Mubarak said. With a collective sigh, the crew dragged themselves back to their feet, picked up their saws and Pulaskis, and went back to work, just as they'd done countless times throughout the summer.

CHAPTER 2

IGNITION

Shortly after winning their tense battle to protect the homes at Adams Lake, Healey's incident management team finished its fourteen-day shift. Like front line fire crews, British Columbia's IMTs work in two-week rotations, stacking many long days in a row before taking a scant few days off to reset, then going back again. The gruelling work schedule can leave many rookies and veterans alike burnt out by the end of even average seasons. Before IMT 5 took their much-deserved break in mid-August, Healey turned over responsibility for the fire to an Australian IMT that—like thousands of other foreign firefighters that year—had been flown in to help plug the growing gaps in firefighting resources as the unrelenting 2023 season stretched on.

The Australian IMT had their work cut out for them. They would oversee two weeks of hard firefighting, first mopping up around the cabins in Dorian Bay and Woolford Point, then chasing the fire as it climbed the mountainsides above the lakeshore homes. The plan was to bring in more heavy equipment and build a layered defence with kilometres of control lines across those mountains and above the Adams Plateau, a high flat ridge that overlooks a small collection of towns along the shores of Shuswap Lake: Lee Creek, Scotch Creek, Meadow Creek and Celista.

Before handing control over to the Australians, Healey and his IMT began sketching out the basics of a fallback: an ambitious planned ignition they hoped they wouldn't have to use. It called for nearly 10 kilometres of high-severity fire following the northern edge of a powerline corridor, stretching west to east, from the top of the Nikwikwaia Creek valley and along the Adams Plateau, before swinging around to the north, and finally ending in the Scotch Creek valley. The ignition would be a last-ditch Hail Mary in case the dozer guards that the Australian IMT was building failed. If it came to it, burning up a huge strip of forest above the powerline as quickly as possible would be their last line of defence, hopefully enough to keep the fire from sweeping down off the plateau and into hundreds of cottages and homes strung along the shores of Shuswap Lake. The paperwork for the aerial ignition wasn't complete when Healey's team left for their days off. It was less a fully formed plan and more the broad strokes of an idea that, if it came to it, they could burn off the powerline as a last resort.

Two weeks later, Healey and IMT 5 returned to Adams Lake and a situation that was deteriorating rapidly. The fire had now grown to nearly sixty-six square kilometres and, despite the Australian team's tireless work, had burned through dozer guard after dozer guard, overcoming ground crews' best efforts to stop it. When Healey's team regained command of the battle, the fire had burned itself into a steep valley that cut northeast into the mountains along Nikwikwaia Creek. Weather forecasting predicted a strong cold front moving southeast across the province. The cold front would bring cooler temperatures but also extremely high winds, and it was expected to arrive sometime after midnight on August 17. Modelling by Healey's fire behaviour analyst Ben

Boghean foretold catastrophic damage. Wind speeds topping 50 kilometres an hour would turn even smouldering spot fires into blow torches. After days of high crossover, it was exactly the kind of fire weather that could level whole towns. And Healey and his team had less than three days to come up with a plan.

As they raced against disaster, more winds arrived that night—a theatrical trailer of the weather to come. Hot winds fanned the flames of the Adams Lake fire with unexpected ferocity. The fire raced up out of the Nikwikwaia Creek draw and cruised along the top of the ridgeline with ease, burning across the last of the dozer guards, shocking Boghean. The Adams Lake fire had bared its teeth for the second time in two weeks, defying firefighters' expectations and proving it cared little for what their modelling tools suggested.

With the Prometheus fire modelling system on its last legs, fire behaviour experts in Canada have been searching for new and better ways to accurately predict what a wildfire might do. One of the most sophisticated new tools Boghean had in his arsenal called FireCast, and is produced by an Italian company called Technosylva. It uses algorithmic machine learning and is capable of ingesting far more data than Prometheus, in theory producing more accurate results. California's wildfire agency CAL FIRE has been testing it in recent years and in 2023, they loaned their new program to the BC Wildfire Service. One of Prometheus's limitations is that it produces binary yes-no projections. Based on the data it's fed, it can predict the fire will burn to X coordinates in Y time frame at Z severity, with little to no ability to gauge confidence or model multiple outcome scenarios.

Along with the traditional Prometheus models, FireCast gave Boghean access to probabilistic predictions that rated the likelihood of each potential outcome. The resulting projection is placed on a map. It looked like a rainbow-coloured smoke cloud billowing out across the North Shuswap landscape, each layer in the cloud a different colour representing a different probability of the fire reaching that far. Based on the data Boghean fed it, the Fire-Cast model predicted that when the August 17 cold front winds arrived, if left unchecked, the Adams Lake fire had a 40%–60% likelihood of hitting the town of Lee Creek (a scenario shown as bright yellow on the map) and a 20–40% chance of engulfing nearby Scotch Creek by midnight on August 18 (light blue).

The worst-case scenario, represented by a pink cloud, showed the fire reaching almost as far east as the town of Celista, more than thirteen kilometres from the fire's known perimeter when Boghean ran the model. But even with this far-greater detail, Boghean wasn't confident. The surprisingly destructive winds the night of August 15 drove fire growth that exceeded both Prometheus's and FireCast's worst-case predictions by more than 100%. "What our modelling predicted the fire would do in two days, it did in under twenty-four hours," Boghean told me months later, traces of awe still in his voice. There was no telling what it would do when the cold front's higher winds arrived.

WILDLAND FIREFIGHTING COMES DOWN TO A FEW KEY PRINCI-ples: the main one is finding a way to break the fire triangle of heat, oxygen and fuel to prevent a fire from spreading. Unless a wildfire is small and still in its initial attack phase, there is typically no putting it out. There is only containing it and waiting for the inevitable rains or snows to finally extinguish it. Even that doesn't always work. Big wildfires can sometimes burn for years, surviving on burning roots underground throughout the winter, only to re-emerge in the spring.

For most fires that survive beyond the initial attack stage, the weakest side of the triangle is fuel. Deprive a fire of new fuel and you stop it from spreading, forcing it to consume what it's already engulfed until it starves itself to death. That usually means one thing: *Build a box and burn it.*

Knowing they can't extinguish a fire as large as the Adams Lake blaze, firefighters instead attempt to encircle or box it in with planned ignitions. Manuals describing the tools and techniques used to do this can sometimes read like a plagiarized copy of *The Anarchist Cookbook*: entertaining reading for any teenager who ever tried stuffing a few hundred match heads inside a tennis ball just to see what would happen when you threw it (not that fourteen-year-old me would ever have attempted such a thing). But the science behind how they work is fascinating and nuanced. It takes ignition specialists years to learn how to properly gauge fuel, weather and topography conditions to predict when, where and how to light a fire to encourage a planned ignition to burn to where it's supposed to. Once it's lit, there isn't much a crew can do to change the outcome if planning conditions—such as the fuel type, density, wind patterns, topography and more—haven't been accounted for correctly.

In Canada, planned ignitions are started primarily in one of two ways: from the ground or from the air. Ground ignitions are the most labour-intensive, requiring coordinated fire crews hiking through the forest using hand-held torches to drip trails of liquid fire behind them. Drip torches themselves are basic implements: a metal can with handle, a controllable vent valve, and a long tube sprouting out the top with a circular pigtail curl in it. The curl prevents lit fuel from flowing backwards up the spout and into the can. At the end of the spout is a nozzle and a wick to hold the flame. The fuel is a roughly fifty-fifty mixture of gasoline and diesel (the exact ratio is varied according to the ambient air temperature)—a combination that will burn vigorously, but not explosively.

In the US, the ignition tools wildfire crews use are about as varied as the firearms that much of the country seems so obsessed with. Alongside the typical drip torches used in Canada, there are all manner of fuses, flares and torches. There are backpack-mounted propane torches and flame-throwers. One company created what is essentially a paintball gun that launches incendiary ping-pong balls. There are flare-launching handguns that wouldn't look out of place in Dirty Harry's grip, and magazine-fed flare-launching long guns modelled after the infamous AR-15. Several companies have even started testing drone-mounted flame-throwers, drip torches and incendiary ping-pong balls to reach further into the forest without putting crews at risk.

At the end of the day, the creativity of the ignition tools matters much less than what the fire is expected to do once it's lit. As one veteran fire safety trainer told me, with the right planning, fuel conditions and know-how, even the trusty BIC lighter will get the job done.

Different ignition patterns will generate different intensities of fire. Sometimes crews create small interlocking boxes of fire that grow together like the ones Murie, Moylan and Lozano used at Tsah Creek to create large swaths of medium-intensity ground fire. Other times crews will work in strips, dragging parallel fingers of fire across a hillside or following the contours of a road or fire guard to create hotter, more rapidly spreading fire that can reach up to the lower branches of some trees.

Hand ignitions like this are used almost daily on the fire line for everything from burning out small islands of fuel left behind by the main fire to cleaning up control lines and achieving a nice consistent black on the fire side of a guard. Small-scale hand ignitions don't need complex written plans or detailed coordination and can usually be overseen by an experienced crew leader working in consultation with their division supervisor or incident commander. Aerial ignitions, on the other hand, are a more complicated beast.

Large-scale planned ignitions, usually delivered from the air, serve different purposes than the more routine hand ignitions used on the ground. Most often, the goal is to consume a large swath of forest fuel quickly and to do so under favourable conditions—rather than waiting for an approaching wildfire to burn it under unfavourable ones. For example, say a wildfire is threatening a town ten kilometres away. Between the fire and the town are several large stands of beetle-killed lodgepole pine. Between the dead pines and the town is a moderately sized river. If the fire were to reach those stands while the winds are blowing towards the town, they could supercharge the fire behaviour and launch volleys of flaming embers towards the town itself, easily vaulting over the river.

On this imaginary day, the weather forecast says winds will continue to blow towards the fire, preventing it from growing towards the town. At the same time, an approaching cold front is expected to arrive in forty-eight hours. It will drive high winds and the fire in the other direction, right towards the stands of dead pines and the town beyond. With the river as a natural fire break to burn off from, choosing to burn up the beetle-killed pines while the winds are favourable could greatly reduce the number of embers available when the fire eventually pushes through them in the coming days, potentially saving the town, or at least reducing the potential damage.

This is, roughly speaking, what Healey and IMT 5 hoped their fallback aerial ignition would accomplish. But they were racing an unforgiving clock. Inside the command post, the team turned their considerable expertise towards figuring out how they could pull it off before the windstorm arrived on August 17. For some ignitions, the goal is to burn up as much surface-level fuel as possible, without significantly increasing the fire's overall behaviour or severity. For this, many fire services use what's officially called a

Plastic Sphere Dispenser (PSD), a name that unironically describes precisely what it does: It drops small plastic spheres, roughly the size of ping-pong balls. Inside each sphere is a granular substance called potassium permanganate. The balls are loaded into a hopper, with a feed tube that hangs out the door of a helicopter. Moments before being released, each ball is injected with ethylene glycol (better known as antifreeze) before rolling down the chute and falling to the ground below.

The genius of PSD balls is the time it takes for the chemical reaction between the potassium and the glycol to kick off—roughly thirty seconds—enough time for the balls to tumble down through the leaves and branches of the tree canopy to the forest floor before bursting into little jets of flame. The resulting fires form small circles that grow outwards at relatively low intensity until they merge with their neighbours. The rate at which that happens, and how spread out the scattered PSD balls are to begin with, helps determine the resulting fire behaviour. By varying the rate at which the balls fall out of the machine, and the speed the helicopter is flying, a talented ignitions specialist can choose the severity and relative flavour of the fire by either dumping balls together in a tight pattern or sprinkling them lightly across the landscape: the firefighter equivalent of seasoning a steak like a Michelin chef.

Using a PSD machine can be very precise but relatively slow, and time was a luxury Healey's team could not afford. In order for their plan to work, they would have to carry it out fast, with enough high-severity fire to burn up as much fuel as possible. They needed a more aggressive option. Luckily, they had one.

Helicopter drip torches, or heli-torches for short, are the other primary aerial ignition tool. They are exactly what they sound like: drip torches of Goliath proportions, slung on a line beneath a helicopter to drip liquid fire onto the treetops. But unlike a handheld drip torch, heli-torches have one key distinguishing feature:

Instead of a purely liquid gasoline-diesel fuel mix, they drip what is essentially napalm. The mixture is a thick, gelatinous substance designed to stick to whatever it hits, which helps keep it in the crowns of trees. Fires lit with a heli-torch are much more intense than those delivered via PSD machine. They are perfect for creating high-intensity, fast-moving crown fires. If a PSD machine is like sprinkling ground pepper on your filet mignon, a heli-torch is like smothering it in ghost-pepper hot sauce.

Planned ignitions can create their own localized wind pattens that can be used to help draw or steer a wildfire in a particular direction. One of the most common uses is to—paradoxically—bring the fire closer to firefighters. A fire burning in steep, inaccessible terrain is often unsafe for crews to attack directly. Using an ignition to bring it down from the mountains onto a shallower slope allows crews to fight it directly, knocking down the active fire front while leaving behind little unburnt fuel on the slopes above that's capable of reigniting.

Because ignitions—like all fires—draw their energy from oxygen, the bigger the fire, the more oxygen it pulls in. Ignition specialists can use this to their advantage as well, sometimes intentionally building a large, hot fire somewhere away from a town, say, on a nearby empty mountainside. If the planned ignition is big enough, the air it sucks in to feed itself creates winds that can act on the encroaching wildfire, pulling or steering it in a new direction and away from the homes or other values like timber or infrastructure that firefighters are trying to protect.

Talented ignitions specialists can even use planned ignitions to create towering smoke columns that have their own tactical advantages. Say the hot August sun is shining down, drying and preheating the forest fuels between a fire and firefighting crews. Winds are calm and expected to remain so for the next few days. Sending a column of smoke into the sky to block the sun from an area that's likely to burn anyway can help shade the area where crews need to work, reducing the temperature and fire risk. The challenge with all of this, of course, is that a planned ignition is betting on a fire to act as firefighters want it to—and even an ignition that has the luxury of perfect planning is still reliant on a fickle, sometimes unstoppable force.

ON THE MORNING OF AUGUST 16, AS BOGHEAN PORED OVER the modelling, it was evident they were running out of both time and options. They faced a difficult choice. Their plan for the aerial ignition was not complete when their team ended their two-week rotation, and they would be hard-pressed to finish it in the time they had left before the August 17 windstorm arrived as predicted. While BC's firefighting policies say small-scale ground ignitions can be conducted at the discretion of experienced crew leaders, complex aerial ignitions require extensive documentation, including detailed mission plans, crew briefings and the careful coordination of ground and air resources to keep everyone safe. They also require a certified ignitions specialist to oversee the operation, but IMT 5 did not have one.

The conservative call would have been stand down, scrap plans for the aerial ignition and have crews do what they could from the ground until it became too dangerous. But the modelling was clear: What was coming had the potential to wipe out homes, cabins, restaurants and campgrounds along a stretch of cherished BC cottage country. From the perspective of Healey and the IMT, doing nothing meant abandoning those lakeside villages to their fate. It also seemed an awful lot like admitting defeat, something wildland firefighters are hard-wired not to do.

The incomplete burn plan called for an ambitious twenty-six-square-kilometre aerial ignition along the roughly ten kilometres of powerline corridor above the towns of Lee Creek, Scotch Creek and Celista. The hope was to consume enough of the forest fuel to create a blackened buffer between the main fire to the northwest and the towns below to the southeast. It was a gamble, a choice between two bad options. Healey's team knew they could do nothing and let the fire crash over homes across the region on its own, or they could try a risky backburn, betting that under the cold front's winds any damage created by the backburn itself would still be less than the existing fire's unmitigated wrath. Even if the backburn caused some damage, they reasoned, it had the potential to save much more. They opted to burn. And so, for the first time in his thirty-two-year career, Healey proceeded without written plans or a certified ignitions specialist.

COMPARED TO ITS CONTEMPORARIES, CANADA HAS SURPRISingly few wildland firefighters. Americans use a wildfire incident command system virtually identical to Canada's, and the overall system functions roughly the same way. The US Forest Service has around 11,300 wildland firefighters, and most states have their own service as well. California, with a population similar in size to Canada's, has more than 27,000 firefighters. In the US, it's common for hundreds or even thousands of firefighters to be assigned to a single fire a fraction of the size that blazes can reach north of the 49th parallel. The cost to hire that many firefighters would be enormous within an economy the size of Canada's, but there is another model.

Australia uses an entirely different system. Major urban centres have their own municipal fire departments, but rural towns rely almost entirely on volunteer brigades that handle wildfires (which Aussies call bushfires). Each town gets an offroad-capable fire engine, protective gear and training from the state. But the labour is all volunteer, and many towns fundraise themselves to purchase additional fire engines and other specialized wildfire equipment. The state of Victoria alone has fifty-one thousand volunteer bushfire brigade members. These rural brigades also respond to far more than just wildfires. They are their towns' first responders for practically everything, including medical calls and car crashes. All of this is supplemented by a much smaller professional bushfire service run by each Australian state, similar to the BC or Alberta Wildfire services. These are home to the true specialists, the burn bosses, fire behaviour experts and complex incident commanders. Like the US, Australia has the ability to put tens of thousands of firefighters in the field at a time and has had to. The 2020 bushfire season in Australia was one of the most devastating in the country's history and taxed even its robust firefighting system to near collapse.

Canada, meanwhile, has far fewer firefighters. They are typically broken down by type, with Type 1 being the most highly trained. These are the IA and unit crew firefighters that belong to provincial wildfire agencies. They are often supplemented by the less highly trained Type 2 and Type 3 crews. Parks Canada, for example, trains most of its trail maintenance crews to a Type 2 standard, which involves less than a week of hands-on training. In BC and Alberta, which have by far the largest wildland firefighting services, Type 2 and Type 3 crews are generally contractors brought in for mop-up and other less urgent work.

Depending on the year, British Columbia has somewhere around 1,000 Type 1 firefighters available, plus additional support staff, experts, helicopter pilots, and others. The whole service comprises roughly 2,000 people. Adding in the available lesser-trained crews, BC has around 1,500 front-line firefighters in a given year. All told, there are only around 3,000 of the highest-trained firefighters in all of Canada, and an additional 3,000 lesser-trained crews combined. Add to that 106 air tankers and 64 helicopters, and you've captured the bulk of Canada's wildland firefighting resources. The only thing missing is the heavy equipment like bulldozers, skidders and water tanker trucks that are all contracted from private industry.

It gets more concerning still. As fire seasons across the globe get longer, hotter and more dangerous, wildfire agencies are struggling to retain experienced firefighters and attract new ones. Canada is far from alone in this, but with so few firefighters to begin with, the bench depth can become pretty shallow pretty fast, especially considering what wildland firefighters get paid.

In BC—the highest paying jurisdiction in Canada—a frontline firefighter's salary starts at $27 an hour, far less than the government bureaucrats who sign their paycheques, earning six figures for a job that does not require them to sleep in a tent for fourteen days at a time, and isn't likely to give them respiratory issues, cancer or leave them with complex PTSD. In Alberta, it's even worse, with a starting wage of only $22 an hour. Some baristas make more than that.

In BC it's not unheard of for some crews to see half their firefighters leave in a single year. During the 2023 season, 80% of front-line crew members had less than twelve months' experience. Almost 60% of their crew leaders were in their first season as leaders, promoted early as veterans grew frustrated and left.

Riel Allain was one of those veterans. He spent six years fighting wildfires for the BC government, and two of those years jumping out of airplanes to do it. To say he has a high tolerance for risk would be a significant understatement. But in 2021, amid what was at the time the worst wildfire season in BC history, he landed on a fire line to a startling discovery. He didn't see the depth and experience of a vibrant, talented wildfire service. He saw rookies—too many of them. It seemed suddenly unambiguous to him that the fire service was stretched beyond its capacity, facing a collision of separate but mutually reinforcing crises.

Wildland firefighting is an inherently dangerous job, something acknowledged by everyone in British Columbia who dons a red firefighter's shirt. The only way to eliminate risks completely would be to simply not fight wildfires at all. Instead, crews use a layered approach to risk management, envisioning a block of Swiss cheese. Each layer is expected to have holes in it. Rookie crew members aren't expected to be perfect. Radio contact in steep gullies might fail. Temperamental winds can blow up the fire on your line. Standard operating procedure can't account for every possible scenario, but safety is maintained as long as these holes don't line up with each other. You can have a crew of relative rookies as long as their leaders and supervisors are experienced and not overwhelmed. You can have a green crew supervisor learning how to shift from grunt work to leadership. You can't safely have both at once, and yet more crews are finding themselves in this situation.

Greener crews require closer supervision, which pulls crew leaders closer to the front line and obscures their view of the larger picture. This in turn puts more pressure on their crew supervisors, increasing the stress of their jobs and driving burnout. When those supervisors look around and take stock of their future career prospects, seeing little but more frustration and a lack of opportunities, many leave. Those empty positions can't be left vacant, so less-experienced firefighters are promoted to fill their boots. This is all happening while the service races to convert itself from a seasonal workforce to a year-round emergency response capability. That means higher level desk-based jobs also have to be filled, pulling even more experience away from the fire line itself.

Ten or fifteen years ago, most crews in BC would have firefighters with a range of experience levels. Rank-and-file members might have three, four, or five years under their belts and only one or two rookies among them. Their crew leaders in turn would often have six or more years of experience, sometimes as much as a decade. It wasn't uncommon for a crew supervisor to have twelve or fifteen years on the fire line. Today a concerning number of crews are approaching 50% rookie status, with two- and three-year veterans thrust into crew leader roles. As holes in each Swiss cheese safety layer get bigger, firefighters face greater risk. Watching all of this unfold, Allain started fearing—and picturing—the worst. He didn't want to be around if, or more likely when, an entire crew got wiped out.

The 2021 season was Allain's last on a BC fire line. He quit and became one of the scores of experienced leaders who have left wildfire services across Canada's west in recent years, frustrated with low rates of pay, a lack of clear career pathways and—most importantly—growing fire line safety issues. But before he left, Allain did something extraordinary. He broke ranks and sent a letter directly to the head of the BC Wildfire Service warning that, if nothing changed, he was afraid the service could wind up facing every firefighter's worst nightmare. "A mass casualty event will come to BC," Allain wrote in 2021. "A likely contributing factor will be having inexperienced crews and supervisors."

ON A FIRE LIKE ADAMS LAKE, AN INCIDENT COMMANDER LIKE Healey would typically have his pick from a combination of unit crews and initial attack—or IA—crews to deploy. To understand how an IA crew works, think of them like a football team's wide receivers. They move fast and can make big plays, especially when flanking around a new fire. When on duty, an IA crew member must never be more than thirty minutes away from their base so that they can respond quickly. When a new fire is reported, IA crews are the first to go, hitting targets quickly or calling in additional resources if they need them. In BC, roughly 90% of fires that IA crews attack are knocked down or contained before they grow beyond a few hectares in size.

When they arrive at a new start, an IA crew will first scout the area and then start cutting hand line with chainsaws and Pulaskis, trying to get enough of a guard around the fire to hold it in place while they set up pumps and hoses to drown it. The key is speed. They often don't have the time or need for backburns.

But just like wide receivers, they can't be the heroes of every down. Sometimes brute force is needed. When it's clear a fire has dug in for a long, grinding fight, that's when a unit crew is usually brought in. Think of them as linebackers. Operating in twenty-person crews and deploying for fourteen days at a time, they don't have the agility to make fast plays like their IA crew counterparts. The job is to muscle the ball downfield a few yards at a time. They're not trying to extinguish the fire but to eventually contain it. They do this through back-breaking physical labour, building layers of fire guard by hand, or dragging kilometres of firehose along those guards through the forest to help attack a fire's flanks. They'll often deploy to campaign fires that are expected to burn for weeks (or sometimes months). They work in conjunction with heavy equipment like bulldozers to build and reinforce machine guards to box a fire in, and then use hand ignitions to reinforce those lines. Good unit crew members are born with drip torches in their hands.

Coast Zulu was a pretty green crew. A four-person initial attack crew, most were in their first or second season with the BC Wildfire Service. Given the intensity of the 2023 season, rookies like them had already racked up many dozens of days on fires in just a few months. Across average fire seasons, it would usually take a firefighter two or three years to earn that much fire line time, but as 2023 underscored, the scale of what counts as average is rapidly accelerating.

While the Adams Lake fire was far from Coast Zulu's first, their role usually had them pouncing on baby fires of a few hectares or less, not doing the grinding work of long campaign battles that unit crews get assigned. But there weren't enough unit crews to go around, and those who were available were already exhausted from months of steady firefighting. In mid-August, there were roughly 370 front-line firefighters, dozens of pieces of heavy equipment and fifteen aircraft split between the Adams Lake, Bush Creek and Rossmoore Lake wildfires, operating out of a fire camp set up at a nearby airfield, more than four hundred personnel in all. Two hundred and forty-nine of them were assigned to the Rossmoore Lake, eighty-two to Bush Creek, leaving only forty-three firefighters assigned to the Adams Lake. In mid-August at Adams Lake, Coast Zulu found themselves fighting one of the biggest and most dangerous fires in the province, alongside their sister unit Coast Romeo and twenty Brazilian firefighters, with precious little backup.

THE MORNING OF AUGUST 16, HEALEY ASSIGNED THE COAST Zulu crew to contain a spot fire near the powerline that runs across the Adams Plateau above the North Shuswap communities of Lee Creek, Scotch Creek and Calista. It had been burning for days, and dozens of helicopter bucket missions still hadn't knocked it down. They worked the fire hard all day, helping to beat back the flames and get a partial guard in place around it. The next day they returned expecting to finish the job. Yet that afternoon, Coast Zulu got a radio call from command telling them to pull back and take up a holding position on the powerline higher up the mountain.

Nobody had told them the planned ignition was happening. In fact, the first confirmation that many firefighters on the Adams Lake fire got was from a social media post by Healey's information officer announcing the burn to the public. Members of Coast Zulu learned that an aerial ignition was happening when they saw a helicopter flying the ridgeline above them, dripping flaming jelly onto the treetops from a bright yellow tank slung beneath it. Until then, the plan had been only a rumour, so far as the crew were concerned. They'd heard chatter about a rough idea to burn a twenty-six thousand-hectare strip, but no one had given them any details, much less told them they'd be directly involved.

Sometime around 6 p.m., as they watched the helicopters flying overhead, a branch director from command arrived with news for Coast Zulu: the job of cleaning up the ignition area along the powerline would fall to them and their sister unit, Coast Romeo. Their task was to burn the southern edges of the ignition area along the powerline itself by hand, because the heli-torch couldn't get close enough to the high-tension powerlines. They'd be working in concert with twenty Brazilian firefighters who had limited experience fighting fires in a British Columbian landscape, and only two of whom spoke English.

There was a lot of information coming in fast from multiple directions, like a high-stakes game of telephone. It wasn't just Coast Zulu that had been left out of the loop. In the rush to execute the operation, multiple senior leaders also found themselves learning about it for the first time as trees nearby roared with fire. What happened next would leave veteran firefighters in disbelief.

One of the most important safety procedures before fire crews undertake a new assignment is what's called the LACES briefing. LACES stands for Lookout, Anchor Point, Communication, Escape Routes (always two of them) and Safety Zones. It's a routine checklist to make sure everyone knows who is watching their backs, what the lay of the land is, what radio channels to be on and where to pull back to if things go haywire. It's so routine that it usually takes only a few minutes. I even have the acronym as a sticker on the water bottle I carry at fires so I don't embarrass myself if a firefighter asks what the LACES are for the day. Several veteran firefighters have told me they'd "never set foot on a fire line without getting a LACES briefing."

Coast Zulu and the Brazilians didn't get an adequate LACES briefing. The window of safety identified by the weather forecasts was already closing rapidly, but they couldn't just call off the ignition. They were already committed, despite the forecasted cold front now just mere hours away. If Coast Zulu had gotten a LACES briefing, chances are someone would have noticed that they had no adequate lookout and no identified safety zones. They might have realized that the language barrier significantly compromised communications with the Brazilian crews. Someone may have realized they had only one escape route, and it involved driving nine kilometres down a hastily plowed access road, which was so rough that pickup trucks couldn't drive any faster than a walking pace, and it was threatened on both sides by fire. Instead, they got a short briefing over the radio describing the basics of their task and told to get after it.

On top of all of this, Coast Zulu had never conducted ignition operations before. Several of them had never held a drip torch on a real fire. As the risks mounted, Coast Zulu's crew leader tried to tell his superiors that he didn't feel comfortable with the mission: "I've never done a large burn before," he told them. Despite these and other red flags, his crew was pressured into proceeding anyway. Despite their reservations, Coast Zulu agreed to go along with the plan, having faith that appropriate planning had been done, even if it wasn't being clearly communicated to them.

The atmosphere in the cab of Coast Zulu's truck was tense as they crawled towards their objective. The treeline above them roared with fire from the heli-torch, while below them the spot fire they'd been fighting to contain earlier in the day threatened their only escape route. They needed to get the burn done as quickly as possible.

Recognizing Coast Zulu's lack of experience, a senior firefighter called a task force leader stepped in to help. He organized a short tutorial on the use of drip torches and guided the rookie firefighters in which drip patterns to use and how best to stay safe while they did it. His mentorship made a significant difference, but things proceeded slowly as the crew adjusted to using drip torches, adding yet more delays to the operation, which was already behind schedule. It was 7 p.m. by the time ground crews got going with their burn task, nearly two hours after the declared safe window to burn had closed.

The work was incredibly slow going. Snaking down the middle of the wide hydro corridor was a thin fire guard that a bulldozer had plowed out only the day before. There was perhaps a few dozen metres between the guard and the forest itself. As Coast Zulu continued pressing east, Coast Romeo worked their way towards the west, each crew burning off from the guard, letting the winds push their flames across piles of brush and brambles and on into the trees. From the air, it looked like a giant had used charcoal to shade in the area between the snaking fire guard and the forest. Above them, fire from the aerial ignition showed through the trees, converging into a bright orange glow, illuminating a towering column of grey-and-black smoke from below as it blew away to the northeast. At first everything looked to be going according to plan, the giant's charcoal slowly shading in more and more as the crews worked along their stretches of the hydro line.

At around 7:30 p.m., the first of two localized wind gusts spooked crews on the ground, momentarily shrouding them in smoke and casting embers across their lines. Coast Zulu's crew leader pulled out his phone and started searching a GPS map for a backup escape route, just in case.

About fifteen minutes later, another, stronger wind shift blew downslope across the entire ignition area, scattering embers and starting spot fires in many areas. Chaos was spreading fast. The Brazilians had been spread out in strategic locations along the length of the burn area, to beat back any flames or embers that managed to jump across the fire guard onto the green side of the line. But no holding operation can withstand a full-blown ember storm. With knots tightening in their stomachs, the Coast Zulu crew realized that they and all twenty Brazilians were about to be cut off and overrun by fire.

CHAPTER 3

TRAPPED

AUGUST 17, 4 P.M., THE POWERLINE

When helicopter coordinator Leanne Ingham first heard the planned ignition was going ahead, she was already airborne in a helicopter, flying over the fire and helping direct other aircraft. As the officer in charge of organizing and coordinating all helicopters in the airspace, Ingham was high on the list of senior staff who—under normal circumstances—would have been briefed on the plan hours or days before it began. Now, along with coordinating several aircraft, she would also have to function as best she could as the ground team's only lookout. Typically, ground crews relied on other experienced firefighters on the ground to be their lookouts, but Coast Zulu had none. Ingham knew she wouldn't be an effective replacement from her position high above, but she was all they had.

Ingham was surprised—and concerned—that such an ambitious burn appeared to be going ahead with so little planning. She wasn't alone. From the air, she watched the start of the burn unfold shortly after 4 p.m. Weather forecasting showed that, by 5 p.m., local winds were expected to become erratic, giving firefighters less than an hour to get the burn completed safely. The heli-torch finished its runs just after 5 p.m. and at first, all seemed to be going according to plan. From the air, Ingham could see fire from the ignition spreading northwards, away from the powerline, just as it was supposed to.

But on the ground problems were already mounting. The Brazilian crews were short some equipment. There wasn't enough burn fuel for Coast Zulu's ground ignition work, and what little there was had been mixed improperly. It was too rich—too much gasoline. Some nearby heavy equipment operators managed to scrounge enough diesel to balance it out, leaving the burn team with barely enough to get the job done. It took more time still for the task force leader to help the rookie crew get the hang of burning with the high pressure task. It was almost an hour before Coast Zulu's ground ignition finally got underway. The ground crews had many kilometres to cover, working along a road so rough it was barely passable for trucks.

Hovering above, Ingham watched the developing situation with a sense of disbelief. The slow pace at which Coast Zulu's burn was progressing worried her, in part because her own flight window was closing. She was unable to fly safely at night and the sun would set shortly after 8 p.m., forcing her to return to base long before the operation was finished. She knew she'd be leaving crews on the ground with no lookout to watch their backs at all. Ingham took photos and video as her helicopter flew, documenting how slowly the work was going. In the footage, the ground crews' trucks appeared as tiny dots of white against the dark ground. Flames glowed orange above them as thick smoke boiled into the sky.

From her vantage point high above, Ingham saw the first gust of wind hit. She watched as it cast embers from the ignition above the powerline into the forest below, and she knew spot fires were likely. Just as Coast Zulu's crew leader had done on the ground, she pulled out her phone and started searching her mapping apps for a secondary escape route—one that should have been planned and identified hours earlier. Roughly fifteen minutes later, she watched as the second, stronger wind shift blew downslope across the entire ignition area, scattering embers and starting spot fires all across the ground crews' lines. From her vantage point above, Ingham saw that the crews' only known escape route was about to be cut off.

Overhead, Ingham worked the radio and simultaneously fired off text messages and photos to other senior staff, showing them in real time what she was seeing. By 8 p.m., spot fires were igniting in heavy timber south of the powerline, pulling the main fire along with it. The smoke column, which had spread high and north all afternoon, now billowed darkly southeast, bending over the ignition line and pushing towards the towns, completely engulfing the firefighters on the ground.

By 8:25 p.m., chaos was spreading. There were spot fires established well south of the powerline and the fire burned freely in the Scotch Creek valley. Two minutes later, the fire had consumed a wide section of the powerline corridor and began spreading vigorously. With the situation deteriorating rapidly, command ordered everyone off the line. But as the crews prepared to evacuate, the fire cut across their only known route to safety.

Frantically, Ingham scanned her maps again. The ground crews desperately needed a new way out. Luckily, she spotted one. A spur road cut across the mountainside south of the firefighters and eventually joined with the Lee Creek forest service road. But to reach it, they would have to drive off-road, navigating around spreading spot fires and over ditches. It would be a pell-mell race to reach safety before the fire caught up to them.

In the chaos, Coast Zulu's crew leader realized that some of the Brazilian firefighters were not receiving the message to evacuate. Rather than head for the newly identified escape route, the Brazilians were driving the wrong way back towards the now-escaping fire itself. One Coast Zulu crew member jumped out of their IA truck and sprinted ahead, down the rough road, through swirling smoke and embers. They knew that, at this point, they could run faster than their trucks could drive. They flagged down the Brazilian crews and passed the message to evacuate. Then they sprinted back to their truck.

Finally, with a convoy of trucks organized, firefighters began making their way back towards their evacuation point, only to find it cut off by flames. To reach the secondary evacuation route, they were forced to drive off-road through bushes and trees. As they drove over a steep embankment, one truck kicked up onto two wheels, nearly rolling over. The convoy passed so close to burning slash piles and spot fires that some of the trucks' plastic components started to melt. Eventually the ground crews emerged from the smoke and reached the safety of the Lee Creek logging road, to everyone's immense relief.

But there was a problem: Not everyone had made it out. From her vantage point high above, Ingham could see what those on the ground couldn't: In the chaos of the retreat, a truck full of Brazilian firefighters had been left behind. Now the fire had trapped them.

At 8:27 p.m., a voice came over the radio into Ingham's ears. It was Shane Derhousoff, the heavy equipment branch director, planning to drive into the fire and attempt to rescue the trapped Brazilians. Unable to raise him on the radio directly, Ingham texted him frantically. "Shane, do not go in there," she wrote.

She saw what he could not: the fire had crossed in multiple spots between Derhousoff and the crew. The forest service road was impassable, and he risked becoming trapped himself.

Ingham urged the trapped Brazilians to abandon their truck and attempt to hike out between the spreading spot fires. But she got no response on the radio. Hovering above, she watched as the flames and smoke bore down on them—their truck now only a barely visible spec of white amid a spreading curtain of smoke, ash and flames.

At 8:30 p.m., as Ingham watched the situation on the mountainside deteriorate, a public update went out to area residents describing the backburn as a success, unaware of the frantic retreat taking place on the ground. "Fire in the ignition area will now burn towards guards," the statement said, "while being monitored by crews patrolling along the powerlines"—the same ground crews that were, at that very moment, racing to escape the flames.

Some residents, who watched the ignition unfold from their properties across the valley, received this update as they were watching the fire escape. It appeared to them that officials were asking them to disbelieve what they were seeing with their own eyes.

Meanwhile, Ingham could see from her helicopter that fire was continuing to spread towards the southeast. From the powerline itself, the Adams Plateau drops down towards the Scotch Creek valley, which runs roughly northeast from the lakeshore up into the mountains. Across the valley from the powerline, the landscape kicks back up again in an abrupt ridge. The ridge itself is met at a right angle by the Meadow Creek valley running roughly northwest. In that valley, twenty-three homes and other structures are spread out along Meadow Creek Road. To the north, there are steep mountainsides, and to the south, a gentler rising ridgeline. As Ingham fired off messages to command about the Brazilians' entrapment and the fire, homeowners in Meadow Creek watched the fire grow rapidly towards them. Everywhere Ingham looked, the situation seemed to be growing increasingly dire.

At 8:44 p.m. Ingham sent a series of text messages and photos to Operations Chief Scott Reynolds, showing the entrapment and the fire burning across the powerline.

"It's blown over the powerlines and moving downslope," Ingham wrote. "If they aren't already, those homes in Meadow Creek need to be evacuated. Blowing hard out of the west and pushing downslope in that direction, slightly south across the line."

"Roger, will talk to Mark right now," Reynolds replied, referring to Healey.

Her text exchanges with Reynolds were the last Ingham made about the fire that night. With darkness spreading and already an hour beyond her required skids-down landing time, she was forced to fly back to base, leaving the Brazilians to an unknown fate. The drive back to the helicopter teams' hotel rooms in Kamloops that night was tense. Ingham and everyone else who witnessed the entrapment were certain they'd just watched the trapped Brazilians die.

UNTIL 2023, CANADA HAD BEEN REMARKABLY LUCKY WHEN IT came to modern wildfire deaths. Between 2000 and 2010, Canada averaged a little more than two deaths per year, many of them either aircraft or vehicle crashes. Since then there have been eight deaths, including three from car crashes, three from aircraft crashes and two Lytton residents who were killed when their town burned down during a horrifyingly fast-moving wildfire in 2021. Deaths among firefighters on the fire line itself are, or at least were, quite rare. That all changed in 2023, starting with the gut-wrenching death of nineteen-year-old firefighter Devyn Gale on the Jordan River wildfire near her hometown of Revelstoke, BC. By the time October rains put an end to the 2023 season, seven more of Gale's colleagues were dead across the country: a decade's worth of lives taken in a single season.

Just nineteen years old and already one of the more reliable firefighters on her crew, Devyn was one of three Gale siblings who wore the red Nomex shirts of the BC Wildfire Service. Their crew mates called them the Gale Force. In the many photos of her that circulated in news outlets and online memorials after her death, she is grinning broadly, exuding a rare confidence and pride for someone barely out of high school.

I was in an incident command post in Northern BC the morning after she was killed. I watched the news roll across the room as firefighters opened the same crushing internal email. It was like everyone took the same gut punch, one after the other. Faces went white, shoulders collapsed. People stood and rushed from the room without speaking.

Like many photojournalists, I've covered my share of traumatic events: car crashes, devastating floods, dead bodies strewn in a residential street, vigils for mass murder victims. I spent years covering the collision of rising homelessness, unaffordable housing and Canada's toxic drug crisis from its epicenter in Vancouver, and it's unnerving how many photo subjects or story sources I've known who have died. All too often, the portraits I made of them in life have a way of becoming their memorial posters. For photojournalists, the camera often acts as a sort of shield or filter, separating and protecting us a little from the sometimes-ugly things we have to photograph.

Standing on the roadside as Gale's memorial parade passed by, I had not expected to start crying. After all, I didn't know her. I hadn't spent any time on a fire line with her. I'd never spoken to any of her siblings. But something about photographing Gale's memorial in Revelstoke, BC, cut through that filter the camera usually provides. Seeing the pain and grief etched on her crew mates' faces as they carried her empty boots, her uniform and her Pulaski past weeping friends and neighbours hit me in a way I wasn't prepared for. Embarrassed, I kept the camera raised to my eye and stayed in the viewfinder, trying to focus on capturing what I was feeling so that others far away might feel a tiny piece of it too.

It was a beautiful ceremony, and I found it brave of her family to invite the media to cover it. In the Revelstoke high school gymnasium, her siblings, Nolan and Kailyn, stood next to a life-sized photo of Devyn when it was their turn to speak. In the photo, Devyn is beaming with her familiar smile, her arms crossed over her chest with an I-can-do-anything confidence. Before a crowd of hundreds, Kailyn spoke about losing not just her sister, but her best friend. About how Devyn deserved better than she got, just as Adam Yeadon did, another firefighter who was killed just days after Devyn on a fire line in the Northwest Territories. Through tears, she challenged the audience to consider whether these deaths were more than just tragic accidents. Maybe Devyn's final gift was to wake us all to a reality we've been avoiding for too long, and a call to action on the climate crisis that firefighters like her family are on the front lines of. Are we really okay with devastating wildfire seasons one after another, with record-breaking temperatures year after year? The world, Kailyn added, wasn't playing fair. "Devyn's favourite tree was the western red cedar," she told us. "And on that day, it was a cedar that took her life. And these summers, these temps, these fires won't stop just because we lost a friend, a colleague or a sister."

After the formal ceremony, there was an informal gathering outside the high school in Revelstoke's Queen Elizabeth Park. It was a sea of red Nomex shirts as firefighters hugged and consoled each other over cookies and soft drinks. Talk quickly turned to the rest of the season, what it might bring and who would be deployed to where. I made awkward small talk and offered condolences to whomever I could, painfully aware that I was not a member of this close-knit community. As we chatted, a veteran firefighter tapped me on the shoulder and pointed to a mountainside west of town.

A fresh column of dark grey smoke was twisting out of the forest into the sky.

THE AFTERMATH OF GALE'S DEATH LED TO A RECKONING WITHIN the BC Wildfire Service, with both internal and external investigations into what happened. While the impact of her death was immediate and obvious, the circumstances that led to it were more opaque. A complex Gordian knot of factors, many of them systemic, helped put her and her crew on that fire that day. In social sciences research, experts call what is happening in the wildland fire world a polycrisis—the collision of many discrete crises at once, each one bad enough on its own, but simultaneously reinforcing the others, creating a vortex of problems. They are so ensnarling that it can feel like the ground beneath a person's feet is shifting even as they try to get enough footing to tackle one part of the problem.

Investigation documents laid out in firefighters' own words serious internal worries that the BC Wildfire Service—considered by many to be the best in the country—is not prepared to face the increasingly common brutal seasons that twenty-first-century megafires have created. "We're not fighting wildfires anymore," one veteran firefighter told me. "We're responding to natural disasters, and we're not prepared for it." In many ways, the sequence of events leading to Devyn Gale's death began weeks and even years before the tragedy itself.

SPRING IS USUALLY A BUSY TIME FOR WILDLAND FIREFIGHTERS in Western Canada. British Columbia is no different. At the Columbia unit crew's Revelstoke fire base in March and April of 2023, the regional offices were dealing with chronic staffing shortages, including several positions that had been left unfilled for almost two years, including a deputy fire centre manager—one of the most senior jobs in the zone, responsible for helping oversee firefighting operations across the region. The trickle-down impact of these vacancies meant that subordinate fire officers and technicians were forced to juggle multiple jobs while also filling their bosses' chairs. The region had been approved for a new fire base, which was already under construction. On top of covering their additional duties, many staff were also busily trying to box everything up and prepare it for the move. Their office space had been cut in half. In short, things were chaotic and had been for a while.

"This led to the zone being understaffed and overworked," an internal report said. Due to a lack of experience, injuries and a 50% turnover among Columbia's initial attack firefighters, the base was short one whole crew—operating with only four instead of its usual five IA crews. Further complicating all of this was the service-wide effort to convert the BC Wildfire Service from a seasonal workforce focused on wildfires to a 365-day all-hazards emergency response agency, one of many priorities set by government to help the province adapt to the pressures of climate change. Amid all the challenges, the zone's acting wildfire officer cancelled a spring refresher training program called the Danger Tree Assessor course so they could attend a leadership meeting instead. The DTA course, where firefighters learn critical skills like how to catalogue and identify all the trees near a work site, plus how to properly assess which ones might pose a safety risk to crews, was never rescheduled.

The shift to 365-readiness has broad support among the public, and among firefighters. It's seen as a critical and overdue effort to prepare BC to face the kind of natural disasters climate change will continue to make more common, while simultaneously providing pathways for firefighters to make their jobs into full-time professional careers, not just summer employment. But it's also a Herculean effort that many firefighters say is being rushed for political reasons in ways that actually undermine the fire service rather than strengthen it.

In the weeks before responding to the Jordan River fire, members of Gale's crew had been dispatched to the massive Donnie

Creek wildfire burning in Northern BC. By mid-July, this monster had grown into the largest wildfire in Canadian history. It would go on to scorch an area as large as Prince Edward Island. While they were there, her crew was involved in a fire entrapment that left at least one firefighter seriously burned.

The morning of the Jordan River fire, firefighters were sitting in a critical incident stress management session designed to help them process the entrapment, but it appears they left with more questions than answers. "The incident up at the Donnie was a shock," one firefighter said. "Learning that someone was maybe going to get skin grafts didn't seem normal. I had a lot of questions about what happened and what went wrong." At least one other firefighter had reservations about being deployed again so soon after such a close call. "Should there not be some downtime between getting burnt over and getting sent back out there?"

Unlike their US counterparts, firefighters in Canada are not equipped with emergency fire shelters. These flimsy-looking sheets of woven aluminum foil, silica and fibreglass, which American firefighters carry in pouches on their belts or line packs, look like a combination of a camping bivy sack and the kind of silvery foil blankets you see at marathon race finish lines. They are capable of reflecting radiant heat of up to 260 degrees Celsius, and they trap a vital pocket of breathable air around the sheltering firefighter. As a last resort when they're about to be overrun by fire (called a *burnover*), firefighters can crawl inside them and boost their chances of survival.

Canadian firefighters do not carry them because burnovers are not supposed to happen in Canada. Instead, fire crews rely on a different, more conservative set of tactics to avoid being in a fire entrapment in the first place, more often fighting large or dangerous fires indirectly with fire guards and planned ignitions rather than attacking them directly as often as US crews might. But that doesn't always stop burnovers from happening, and as climate change continues to supercharge our wildfires, the risk of explosive fire growth suddenly overtaking crews, like it did the Brazilians at Adams Lake, continues to rise.

The call for the Jordan River fire came from Revelstoke on the afternoon of July 12, 2023. A small plume of white smoke was visible in the rugged mountainside about five kilometres northwest of town. An officer from the Columbia Zone was dispatched in a helicopter to recon the fire and submit a report and, once overhead, requested one IA crew with a certified tree faller and some still wells: portable fabric water tanks to feed pumps and hoses when no natural water source is available. The fire was roughly ten metres square and burning at rank 2, the second-lowest on the six-tier fire behaviour scale. It looked like an easy target and a relatively quick job.

Access to the fire, however, was a significant challenge. It was burning on the side of a steep valley covered in decadent heavy timber—giant cedars and Douglas firs. The only possible helicopter landing site was higher up the valley, beyond a narrow gully that crews would have to hike across. Despite only being six hundred metres away, the terrain was so rugged the hike would take the crew more than forty-five minutes, even without any gear on their backs. Six firefighters went out the first day to better size up the fire on the ground. They scouted around the fire's perimeter, a challenge in steep terrain shot through with slide alder and thorn-covered devil's club. Beyond an initial assessment, including the need for more certified fallers to deal with dangerous trees, the crew spent little time on the fire that day. They were all back at base by 8:30 p.m., with plans to tackle it again at 7 a.m. the next morning.

The next day, firefighters returned, this time with a plan of attack. The fire had shown little overnight growth and was still smouldering away between rank 1 and rank 2. At 11:40 a.m., the crew leader and incident commander—who investigative documents identify as "Blake"*—radioed in their plan. By the end of the day, they aimed to cut a fuel-free line around the fire, clear a second heli-spot closer to the fire itself, and fell all the danger trees nearby so crews could work safely around the fire. To help speed things along, several firefighters, including Devyn's brother Nolan, spent the morning helping the helicopter sling-load their gear across the narrow gulley rather than risk lugging it across by hand. By now the fire was burning hot enough that still wells weren't going to provide enough water to fully extinguish it. The crew would need a nearly one-thousand-litre water bladder flown in that could be resupplied by helicopter bucket drops. That meant more saw work to clear the bladder site, plus more cutting to create an access trail from the original heli-spot.

At 12:30 p.m., Devyn and two first-year firefighters identified as Jesse and Carter† were flown in to help. Though it was Devyn's second year as a fully-fledged crew member, she'd also spent a year in high school working on the BC Wildfire Service junior fire crew program. She was a certified bucker (meaning she could cut up fallen logs but not fell trees herself) and a sought-after mentor to rookies on the fire line. With Devyn, Jesse and Carter deployed, there were now nine firefighters working the fire.

Blake, the incident commander, was a crew leader with more than a decade of wildfire experience. She completed a danger tree assessment of the worksite. Any time a potentially dangerous tree is identified, policy dictates it must be either cut down or surrounded by a wide "no work zone" marked with brightly coloured flagging tape. Such assessments are one of the first and most important tasks

---

* The BC Wildfire service calls these documents Facilitated Learning Analyses. They are not findings of fault and do not assign blame. Rather, they are narrative descriptions of the events and circumstances that contributed to an incident, with the goal of learning and avoiding mistakes in the future.

† Everyone except Devyn named in the FLA documents is given a pseudonym to protect their privacy. I've chosen to use those same pseudonyms here.

on a new wildfire worksite, usually followed by a DTF—Danger Tree Falling—something that can only be carried out by certified fallers.

Inside their worksite perimeter, Blake identified a cluster of three cedar trees, one of which had already fallen. Just below it was another cedar more than twenty metres tall with a large widow-maker snagged in its upper branches. Next to that was a third massive trunk, burning from the inside out like a chimney, the top portion of which had already failed and come crashing down overnight. It burned so intensely that the roar from its fire could be heard by firefighters too far away to actually lay eyes on it.

During a lunch break, firefighters discussed the risks posed by the cluster of massive trees. A firefighter identified as Joe told the crew a story about a near miss the summer before with a very similar tree. He'd had to scramble out of the way as a burning old-growth cedar crashed to the ground just metres behind him. The risk that this cedar might also tumble with little warning was unmistakable but, as with most decisions in a wildfire, the calculus was far from simple.

As long as the tree remained standing, the fire inside would spiral harmlessly up into the sky. Dropping a twenty-metre cedar full of flames risked adding more fire on the ground and could effectively double the size of their fire before the crew had a reliable water source set up to contain it. Crew members were split on what to do. Joe was sure the tree would fail and didn't feel comfortable working near it. Devyn suggested a second danger tree assessment be completed. Other crew members said that in their experience cedars were usually stable. In the end, no second DTA was done, and a no work zone was not established. Instead, the crew decided to take a "heads-up" approach. The tree would be left standing until the water bladder and a hose system were set up to contain any fire spread that dropping it to the ground might cause. Everyone got back to work cutting trail around the fire, all working in pairs except for Devyn, who worked her way around the top of the fire on her own, using her chainsaw to cut a fuel-free swath through the dense forest.

Not long after, Jesse heard the loud crack of a tree falling from near where Devyn had been working. Blake called for everyone to radio in their positions, but as their voices crackled over the radio one by one, there was no response from Devyn. Worried, Jesse pushed through the thick, snagging undergrowth towards Devyn's last location. He found her sitting upright, her legs pinned under the tree, face bloodied and her hard hat crushed. Weakened from the fire inside it, the tree had sheered into several giant slabs as it fell. Devyn's crew sprang into action. Jesse yelled that Devyn was injured and started trying to cut up the massive slabs of burning cedar with his chainsaw, trying to free her. But getting her out from under the tree would be only their first problem. Devyn was alive, but barely. Getting her off the mountain fast enough to save her would be much harder.

Fire centre dispatch call logs show a radio transmission at 1:20 p.m. "Medevac—firefighter struck by a tree send hoist extraction." As Jesse worked his saw on one side, Blake joined him and started cutting on the other, both of them ignoring the burns they suffered and the smoke they inhaled as they fought to free their friend. Hearing the distress calls on the radio, another firefighter identified as Logan raced down the hill to help. As he arrived, he and Jesse began to pull Devyn out from under the tree and cut away her burning chainsaw chaps. She was unconscious, and her breaths came in ragged, laboured gasps. As Jesse and Nolan administered first aid, Max took his saw and started felling trees to clear a spot so the helicopter rescue crew would have a safe space to rappel into.

As the radio distress calls went out, helicopter operations tech Hayden was en route back to the Salmon Arm base after deploying an IA crew to another nearby fire. As he prepared to join the extraction mission, he grabbed extra first aid gear and two spare oxygen tanks. He also pulled a firefighter identified as Sasha onto the flight to help. All they knew at that point was that there'd been a tree strike, and it was serious. At 2:14 p.m., as they left the Kamloops zone and crossed into Columbia's airspace, they heard a radio request for an automatic external defibrillator, or AED. The situation was worse than they'd imagined.

At 2:37 p.m., Devyn was loaded into an aerial rescue basket and hoisted into the helicopter, where Hayden and Sasha took over first aid. They kept up their efforts even after landing, helping hospital staff as they worked. Hayden kept up CPR for almost an hour, fighting until the very end to save Devyn's life.

The doctors who treated her at the Revelstoke hospital would later describe the medical care Devyn received from her crew mates as nothing short of heroic. From the moment the tree came down, everything they did was executed flawlessly, and under extreme stress. Everyone involved helped deliver sophisticated trauma medicine far beyond what anyone could have expected of them. Together they gave Devyn the best possible chance they could, and it still wasn't enough.

The impact of Devyn's death was immediate and profound. Finally back at base, Blake collapsed sobbing in her crew's locker room. Others wandered around as if in a daze, "shell-shocked," as one of them later told a government investigator. The incident had unfolded so close to town that other firefighters, including the crew leader who'd been seriously burned at the Donnie Creek fire, could hear the radio traffic as the response unfolded. Sitting on base, there was little they could do but listen with a growing sense of dread.

While the emotional crater Devyn's death left on base was wide, the incident also revealed much deeper gaps within the fire service and left many firefighters questioning its ability to handle the aftermath of traumatic incidents like this. While everyone on base was still coming to grips with Devyn's death, many of them were also asked to keep working. Some had to train staff from an independent workplace safety agency how to exit a hovering helicopter, a tricky and dangerous maneuver in its own right, so they could access and investigate the site where Devyn had been killed. Others had to coordinate counselling, plan dinners and help manage their bosses' schedules, all while a devastating wildfire season raged on around the province. Over the next three days, almost everyone involved was interviewed repeatedly by the RCMP and WorkSafeBC investigators, having to retell (and relive) their traumatic experiences multiple times with little room for breaks, and without being told some of the interviews were not mandatory.

Complicating all of this was the internal pressure many of them felt to get back to fighting fires. They were encouraged to take time off if they needed it, but many worried about how or if they would get paid if they did. "None of us were in any state to do work, but we did it," one of them said.

Some of Devyn's fellow firefighters would later question whether the fire she died on had needed to be fought at all. The Jordan River fire was highly visible from town, and many firefighters speculated that public pressure to suppress visible fires likely influenced the decision to send crews. Compounding things further, front-line firefighters questioned whether senior staff had the operational experience to safely tell the difference between a fire that must be fought urgently and one that could be left to burn on its own. "We end up actioning fires that we don't need to. This fire wasn't going anywhere," one firefighter said.

For a fire like Jordan River, burning in a decadent stand of old-growth trees with no timber harvesting plan and no homes at risk, in an ecosystem not prone to rapid fire growth, it's likely the fire could have been left to its own devices. That's exactly what happened when another lightning-caused fire sparked nearby ten days later. No crews were sent, the fire did its own thing, and it eventually burned itself out at 1,265 hectares, burning to within a kilometre of the Jordan River fire.

Making the call to send firefighters to a wildfire or not is often complicated, informed by far more than whether or not homes might be at risk. Many people, including me, thought the Adams Lake fire would burn harmlessly into the forest, helping to restore balance to the ecosystem and harming no one. Few people foresaw the erratic winds that drove its explosive growth on August 2 and left crews scrambling to defend homes along the lakeshore, just as no one expected how badly the August 17 windstorm would supercharge the fire. But one thing is clear: The compounding impacts of longer fire seasons and more extreme, unpredictable fires are forcing firefighters into increasingly dangerous situations more often. Simple math dictates the eventual tragic outcomes from there.

Along with overwork, burnout, and public pressure to suppress visible fires, many firefighters interviewed after Devyn's death said the high rates of crew turnover were a major contributing factor—not because any of the rookies made a mistake that day, but because of how high turnover adds additional pressure to a system already straining at the seams. "It's like an army of kids out there," one veteran crew leader told me.

Most firefighting can't be taught in a classroom. It has to be learned through experience in the field. This is exactly how the BC Wildfire Service works best: hands-on learning under the supervision of experienced leaders who've been doing it for years. But rising crew turnover makes it difficult for crew leaders and supervisors to maintain a broad enough span of control to keep everyone safe and also teach rookies at the same time. It all adds to the pressure, fraying an already threadbare system. Simple safety checks can get missed, routine briefings skipped.

Of the nine firefighters on the Jordan River fire the day Devyn died, two had more than a decade of firefighting experience. Just a couple years out of high school, Devyn was the next-most experienced. Everyone else was either a new hire or had one relatively slow season under their belts. "Being the only experienced person on a crew is scary," said one crew leader quoted in the documents. "There are not many five- and six-year firefighters," said another.

The BC Wildfire Service has been aware of these challenges for years, and is making significant strides towards addressing them. Central to their efforts is the creation of Canada's first university degree for wildland firefighting at Thompson Rivers University in Kamloops. The goal is to help build wildland firefighting from a summer job for 20-somethings into a viable long-term career. The service has also added regimented nutrition programs, psychological support and athletic therapists to many of their fire camps. They are essentially working to treat firefighters less like temporary labour and more like the high-performance athletes they are.

These efforts are impressive, and in many ways nation-leading. Imagine coming off the fire line to a physio treatment and a healthy meal instead of a paper bag dinner. But building a university program from scratch takes time that even Canada's most robust wildfire service doesn't have. And even if it's successful, all the professional wildland firefighting in the world won't be enough to hold back the monster fires that are now possible.

High turnover not only compromises the effectiveness of individual crews; it can have consequences at wider operational scales as well. Incident commanders, particularly on large fires when they're overseeing dozens or hundreds of firefighters, can't possibly be expected to know the details and real-world capabilities of every crew they send out. To most commanders, a unit crew is a unit crew, a tool to accomplish a job. If a particular incident commander has thirty years of experience, that also means thirty years of built-in expectations about what a unit or initial attack crew can accomplish, and assumptions about their level of experience and ability. A commander might dispatch a crew with a particular mission, not realizing they lack the necessary experience to safely pull it off. Remember the block of Swiss cheese? Any number of things can cause another hole in a single slice: attrition, internal promotion, firefighters leaving the service. As the experience levels of crews and crew leaders begins to decline as a result of these factors, the holes only get bigger, more numerous and more likely to line up. Eventually, and especially in extreme seasons like 2023 when resources are stretched beyond their capacity, that can have catastrophic consequences.

It's how, for example, a relatively green initial attack crew like Coast Zulu ends up having to learn how to use drip torches next to nine kilometres of crown fire from a hastily planned aerial ignition with an inadequate safety briefing and no viable escape route. It's how they end up doing this while also trying to manage twenty foreign firefighters who don't speak their language ahead of a windstorm on one of the most devastating wildfire weekends in BC history.

AUGUST 17, 9:30 P.M., MEADOW CREEK ROAD

"OH, FUCK..."

Standing in his yard at the head of the Meadow Creek valley, Dave Dyck craned his neck and peered into the smoke-dark sky. For a brief moment, the wind had abated, and the sudden quiet caught him off guard. It was haunting, distracting, like the moment when an angry beast pauses to draw in breath. Thousands of embers hung seemingly motionless in the air, suspended in time like so many bright orange fireflies. Then, as suddenly as it had stopped, the beast roared again, bathing the world around Dyck in embers, heat and ash. "Oh fuck!" he said again, this time sounding less like wonderment and more like fear.

Standing in a field of garlic bulbs Dyck gripped his fire hose tighter and went back to spraying water in wide arcs, back and forth, soaking the field to keep the embers from catching. In the hills around him, the whole world appeared to be on fire.

Dyck bought his collection of Meadow Creek properties in 2013 hoping to get back to his Mennonite farming roots. They sit at the valley's highest point, with steep mountainsides rising to his north and south. The road dead-ends about sixty metres past his driveway, the landscape dropping steeply down towards Scotch Creek itself before climbing again up towards the Adams Plateau and the powerline where the aerial ignition was lit. The only way in—and more importantly out—is back down the Meadow Creek Road itself. If that were to be cut off by fire or fallen trees, Dyck and the handful of his neighbours who stayed to face the fire and its cold front driven winds, would be trapped.

As the fire roared up out of Scotch Creek valley, it barrelled towards Dyck with an energy he describes as a dragon, shooting flames more than ninety metres into the air. In the hour after it trapped the five Brazilian firefighters and forced the rest into a chaotic retreat off the mountainside, the fire had burned down into the Scotch Creek valley and climbed the ridgeline up into Meadow Creek. Dyck had watched it the whole time as he prepared to defend his structures, as did other residents with properties scattered along the nearby ridgelines. Now the fire was racing up the hillsides on either side of Dyck's home, roaring through the trees.

Before the fire, Dyck had spent years preparing to defend his properties. Every building had permanent structure protection sprinklers installed—metal ones that can throw water in a twelve-metre arc 360 degrees around the building. Knowing the risks a wildfire posed, Dyck's long-held distrust of government meant he didn't want to be reliant on anyone but himself. He also dug a pond, eighteen by thirty metres, fed by an artesian well, and installed a high-volume pumphouse complete with its own backup generator. It feeds out to a high-pressure firehose with wildland fire fittings, the same kind that firefighters themselves use. Along with that, he has several wildfire-style mobile pumps he can plug in to other creeks in the vicinity.

"Every corner of this property I can get water to," he said. He has several giant fuel tanks, as well as smaller portable ones to run all his generators and pumps. Each of his three houses has its own water reservoir. His latest build, which he planned to flip into his retirement fund, has a twelve-thousand-litre tank. Dyck said he keeps between one and two years' supply of food and fuel on his property at all times. "When shit hits the fan, I'm gonna eat," he told me.

For days before the fire's arrival, he had his water systems set up and running, soaking as much of his property as he could—his garlic fields, the forest, everything he could—knowing it might mean the difference between saving his investments or losing them. He was pumping out so much water—thirty thousand litres a day, he said—that he created a giant dome of moisture and high humidity over the whole area.

As the fire climbed up out of the Scotch Creek valley, looking for a foothold in Meadow Creek, Dyck told me that it ran into the moisture dome he'd built and, like a wave breaking against a rocky outcropping, the fire split in two, racing up the ridgelines to the north and south instead of tearing along the valley's floor. Dyck said he doesn't have perfect recollections of everything from that night, but he does remember the municipal fire department eventually arriving and urging him to leave. Despite the obvious risks, Dyck couldn't bring himself to abandon his land. "I can't let this burn, this is my retirement," he told them.

At around the same time, Celista Fire Chief Roy Phillips was standing eight kilometres south, outside the Scotch Creek fire hall with the captains of two other nearby fire departments. They were discussing what to do if the fire exploded, where to position firefighters and how they might execute an evacuation order if one came down.

The smoke plume from the aerial ignition had now been looming over the town for almost five hours. And for most of that time, it had behaved as Phillips and the others had been told it would: spiralling up and away to the north. To anyone standing on the front pad of the Scotch Creek fire department, the powerline itself is hidden from view behind another ridge closer to town. All Phillips had seen of the burn's first few hours was the towering column of smoke. He had no idea the fire had overrun the powerline and pushed crews into a chaotic race for safety.

But when he saw the first telltale signs of flames flickering on the ridgeline to their north, he could tell the fire had breached its containment lines and was headed south towards them. Phillips was surprised and worried. He and the other fire captains started making calls to the regional district and the wildfire service, but no one was answering.

Immediately, Phillips knew something was amiss and started dispatching his crews around town to keep a lookout for spot fires. There'd been a spot fire in the Meadow Creek area a few days earlier, and Phillips's crews had fought it. He swore it was contained and extinguished before the aerial ignition began, but just to be safe he sent a crew in to check it out.

The rural homes in Meadow Creek are sprinkled along the valley bottom, set well back among the trees. There's no reliable cellphone or radio coverage, so Phillips's crew leader had to drive all the way back out in order to make his report. What he said left Phillips floored. He couldn't believe the fire had already reached homes in the valley, so he headed in to see for himself, driving amid smoke and flames. As he drove deeper, he found local residents already trying to evacuate, and others frantically trying to fight the flames themselves. As the fire moved through, residents started calling and warning each other, reporting the fire's progress as they rushed to gather belongings and get out. As Phillips patrolled the area, knocking on doors to alert homeowners, he found dinner plates with half-eaten meals that were left where they sat. It looked like a scene from a sci-fi movie scene, one where a whole family vanishes in an instant. "It was the eeriest thing I've ever seen," Phillips said. "And they had no notice other than calling each other, saying, 'Get out of the house, now.'"

While many homeowners were either in the midst of fleeing or had already left when Phillips arrived, with no evacuation order in place, others refused to go. Phillips found one man standing next to his home with a garden hose in his hand. "This is too big," Phillips said. "You've got to get out of here. Tonight. And he just said 'I can't. I've got to save my house.'"

"I told him, 'The last thing I want to do is have to come back here and identify a body. It's not worth your life. You need to survive.'"

Though he can't be sure, Dyck was likely one of the people Phillips ran into during his patrols of Meadow Creek, when he tried to get everyone to leave. There was little Phillips could do to convince the man, so he headed back to the road junction at the head of the valley where he could get a good cell signal again. Crews from other fire departments had now arrived as well, but Phillips was forced to make a difficult call.

The winds blowing southeast were pushing the fire right towards them, with the valley walls acting like a funnel. There was active fire on both sides and only one way in or out. All it would take was one downed tree, or a car crash or any number of other mishaps, and the valley could easily become a death trap.

Fearing for his firefighters' safety, Phillips held them at the head of the valley and instead focused on trying to convince the wildfire service and the regional district to issue an evacuation order. And all the while, frantic local residents were coming up to him, yelling at him, demanding to know why he wasn't down there saving their houses.

It was now 10:03 p.m. on August 17, and as more municipal firefighters arrived, Phillips decided to take one crew and head back down the road to recon the situation. To his surprise, many homes had sprinklers already installed by the homeowners themselves. Others had their own wildfire pumps and were busy fighting the fire around their houses. Another hour later, and more than two hours after Ingham had called for one, the regional district and the wildfire service finally managed to issue an evacuation order. Between the efforts of locals like Dyck and significant luck, no homes were lost that night in Meadow Creek. Despite that good fortune, the situation across the North Shuswap was still desperate. Coast Zulu and most of the other firefighters had narrowly escaped the blaze. But five Brazilians were still trapped in their truck, surrounded by fire. The battle for the North Shuswap communities was just beginning.

CHAPTER 4

CHAOS, COURAGE AND CONSPIRACY

AUGUST 18, 12:20 A.M., MEADOW CREEK ROAD

As August 17 turned into August 18, Healey received Fire Chief Phillips's calls to send more firefighters to Meadow Creek. He likely heard the desperation in Phillips's voice, but there was nothing his wildland crews could do to help. Wildland crews aren't trained to fight structure fires, and with the fire burning in steep, unsafe terrain around and above homes on the road, it was not what wildland firefighters call an "achievable objective." But there were other resources Healey could send.

At 2:30 a.m., a structure protection specialist named Jesse[*] woke up to his phone ringing. It was Healey. Strong southerly winds in the Scotch Creek valley had sent the fire into Meadow Creek Road, and Jesse's structure protection task force was being sent in to help, along with fellow structure protection specialist Cameron. The fire had now been burning around homes in the area for almost five hours. Jesse's task force arrived at 3:18 a.m. and linked up with Phillips's firefighters from the Celista Fire Department. Together they fought alongside locals to protect properties along the five-kilometre stretch of road.

By now, the fire had established itself well south of the powerlines, both at the head of the Meadow Creek valley near Dyck's house, and in the hills above Scotch Creek. The catastrophic windstorm that weather forecasting predicted days before had arrived. Winds from the west whipped in at thirty to forty kilometres per hour, sometimes gusting as high as fifty kilometres per hour. Phillips's firefighters knew they would be in a crucible of fire for hours.

They weren't the only ones feeling a heavy sense of foreboding. At 7 a.m., Ingham walked into the fire centre manager's office in Kamloops visibly distressed. She told the fire centre manager about the previous night's entrapment, which she still feared had killed nearly half a dozen people, possibly more. She didn't know that hours before, as they sheltered in their truck amid the blowing embers, flames and smoke, the Brazilians had heard her messages to flee and had chosen to ignore them. Luckily, the Brazilian crew was more experienced than their BC colleagues had realized. Several of them had survived similar entrapment situations while battling fires in their home country. They knew how to keep their heads, and that evening they'd had another idea for surviving the oncoming fire.

---

[*] While it is not possible to know for sure, it's my belief this is a different "Jesse" than the one identified with the same pseudonym in the Devyn Gale FLA documents.

Rather than risk hiking out through the flames, they parked their truck in a wide clearing directly under the powerlines, headlights pointed into the oncoming blaze. They raced to gather burning branches from around their vehicle, dragging them in circles through the tinder-dry grasses and shrubs, burning off any potential fuel before the main fire front arrived. This technique is sometimes called a rescue fire, and in dire situations it can be the last, best chance for fire crews to save their own lives.

With their rescue fire complete, the Brazilians climbed back inside their truck with the engine running, sealed up the windows, and cranked the air conditioning to full, just as their training had taught them. The move was itself a significant gamble, a bet that relied on years of experience and the confidence to hold fast as the fire ran over them. The truck was their best available shelter, but if it started to burn, they couldn't get out and run. They'd be in the middle of the fire itself.

"We're here, in the fight," one of them said. "It was good to be with you guys," added another, trying to keep everyone's spirits up as the world outside the truck was reduced to a hellscape of swirling smoke and flying embers. Bright billows of flame were only ten metres away. Burning debris rattled against the windows like hailstones as the fire swept in around them.

The Brazilians sheltered in the truck for hours. They were ultimately rescued sometime after 11 p.m. by two senior BC firefighters who drove back through smouldering embers and thinning smoke in the now-burned area, bucking fallen trees off the road with their chainsaws. When they reached the stranded crew, the Brazilians were all fast asleep. They were finally escorted back to the base, a four-hundred-person fire camp set up at the Squilax airfield, by about 2:30 a.m., just as Jesse's task force was being spun up to Meadow Creek.

As Phillips and his crews fought to push back the fire spreading around Meadow Creek, the winds continued to pick up, pushing fire south towards town along a fire front nearly thirty kilometres wide. Back at the fire camp, a debrief of sorts was ordered. The chaos of the previous day's backburn had yet to fully sink in. Many people were upset about what had happened—about how close they'd come to death. Still shaken from the night's close call, neither Coast Zulu nor the now-rescued Brazilians were fit to work the fire line that day; they were assigned less stressful camp-related duties instead. As the meeting progressed, another dark column of smoke loomed in the distance, this time on the west side of Adams Lake and directly north of the camp itself. The Bush Creek fire had blown up and was taking the first steps of an enormous twenty-kilometre run south.

By mid-morning, reinforcements started to arrive in the area. The Bighorn and Firestalker unit crews—forty wildland firefighters in total—went to work setting up sprinklers and clearing flammable material from around people's houses along the west shore of Adams Lake as homeowners prepared to evacuate. It was now obvious to everyone that the fire's impact on the region was going to be extreme. In Scotch Creek, locals raced to fuel up their boats and left them sitting at a local marina, keys in the ignitions, ready for a hasty retreat.

Back in Scotch Creek, Dean Acton and his son Mark were hearing reports of the fire advancing towards town. The father-son team runs the town's marina. As the fire began showing itself in the hills north of Scotch Creek, Mark organized a group of local residents with heavy equipment, including an excavator, and started trying to plow a fuel break through the large, open field north of town.

A few hundred metres to the Actons' south, workers raced to set up a series of massive high-volume sprinklers. Twelve of these monster water cannons were arranged along a snaking three-kilometre line fed by thirty-centimetre hoses and powered by two massive 604 horsepower pumps drawing water from the nearby lake. At full capacity, the system in Scotch Creek could shower a combined thirty-five thousand litres of water per minute in twelve interlocking circular arcs almost two hundred metres wide.

As the fire loomed and pressure mounted, the fire services' chain of command started to show cracks. The Bighorn unit crew was getting competing orders from two different division supervisors and one of the helicopter coordinators. The Firestalker unit crew arrived and was assigned to help with structure protection near Holding Road, a strip of houses and properties that runs south from a large sawmill on the southwest shore of Adams Lake. It's connected by ferry to Woolford Point and Dorian Bay, the same neighbourhoods that had barely survived the fire's first big blow-up weeks earlier. From the ferry landing, Holding Road then continues south towards the airfield fire camp. Within minutes, the Firestalker crew ran into frantic locals who were unsure whether to flee or to stay and attempt to fight the fire themselves. No evacuation orders had been issued yet, but it was clear things were getting dire. They didn't have to wait long; the first of those orders finally came at noon, when residents in the communities of Dorian Bay and Woolford were ordered to leave.

By now fire was driving forward on two main fronts—the original front line southeast from the aerial ignition stretched nearly twenty kilometres east from Lee Creek to Celista and beyond. At the same time, the Bush Creek fire was burning a second front more than five kilometres wide, south through the hills above the western shore of Adams Lake, making a run straight for the BC Wildfire Service fire camp at the airfield on Squilax-Anglemont Road. In the coming hours, the two fires would merge into one raging behemoth, as fire raced down the Adams River canyon and exploded out across the bottom of the valley below.

The Bush Creek fire made its first big run towards the fire camp in the same moments the Dorian Bay and Woodford evacuation orders were issued. Flames surged against a series of giant water cannon sprinklers that had been set up around the Adams Lake sawmill—one of the area's shelter-in-place locations. Firefighters use these locations as havens of last resort, a defensible area they can fall back to if they're overtaken by fire and there's no other escape route available. A wall of smoke and flames bore down, with the sprinklers firing full blast to hold back the fire. The winds were now so strong they whipped Adams Lake into white-capped waves more than a metre high.

At 1 p.m., Jesse's task force was ordered out of Meadow Creek. Fire behaviour in the area was becoming too extreme, and they were needed to help defend the sawmill. But that meant reluctantly leaving residents who were still trying to escape and couldn't drive past the approaching flames. Those residents were forced to shelter in an open area at the Shuswap Lake Park, while dozens of others were fleeing in boats across the heaving, wind-driven Shuswap Lake. It also meant Jesse's task force had to drive more than thirty kilometres through a region now riven by fire.

When the task force reached the Holding Road area south of the sawmill, they found fire had already surrounded multiple nearby homes. Jesse watched as waves of fire rolled across the road in front of him, jumped the Adams River with ease, and ran back up the other hillside with a speed and ferocity that left him dumbstruck. It was fire behaviour the likes of which he'd never seen. His team drove with flames and smoke all around them, so thick they struggled to see the road's yellow centre line. It was exactly the kind of scenario he'd always instructed other firefighters *not* to get into. To veteran Operations Chief Scott Reynolds, watching from the incident command post at the fire camp ten kilometres to the south, it looked even worse. It looked like hell coming out of a mountain.

This kind of explosive fire behaviour was taking place all across the North Shuswap, and emergency managers were completely overwhelmed. Nobody could keep up, not Phillips's fire crews in Meadow Creek, nor Jesse's task force, nor the Bighorn and Firestalker unit crews nor the panicked residents along Holding Road, not even the municipality's emergency operations centre. Homes were already burned to ashes by the time official evacuation orders were announced for Celista. In other communities where evacuation orders still had not been given, hundreds of residents were fleeing for their lives.

By 1:30 p.m., the fire was burning towards the Scotch Creek bridge, flames licking along the forest floor through dense smoke mere metres away. The fire now threatened to cut off the only road out for hundreds of people living east of Scotch Creek. Firefighters had been going nearly non-stop for more than fourteen hours. After fighting to defend homes in the Meadow Creek valley the night before, Roy Phillips was running on only about an hour's sleep. So were most of his firefighters. He dispatched them to the bridge anyway and joined them just after 2 p.m.

When he and his crew arrived, members of a BC Wildfire Service structure protection task force were already there. The bridge was surrounded by fire, but the structure protection in place appeared to be holding. Sprinklers had created a broad moisture dome that was—for the moment at least—protecting the bridge. It would be a close fight, but Phillips figured they could save it. But when he began to deploy his structural firefighters, the wildland task force crew told him they were being ordered—against their wishes—to pull back and abandon the bridge. Command feared that if the bridge failed fire crews might get trapped on the wrong side with no other escape routes at all. The task force leader told Phillips to move his firefighters to the far side as well.

"You need to get your guys off of this side of the bridge and back to the other side," he said. "We have rank five fire coming down this mountain now, right at you."

"I have to stay," Phillips replied. "We have to look after our community. You guys go ahead, but we can't go." If the bridge were lost, there'd be no way out for the hundreds of people still trying to flee Scotch Creek and Celista. Despite the risk, Phillips couldn't bring himself to abandon it.

By midday on August 18, Ty Barrett, chief of the Shuswap Fire Department, was called into action. His crews are typically responsible for homes on the south side of Shuswap Lake, but now they'd been pressed into service against the fire as well. Like Phillips, they arrived at the bridge around 2 p.m. Barrett saw the protection sprinklers on the bridge were doing their job, but fire was creeping closer, burning in from both sides of the road. Cars full of evacuees were still fleeing across the bridge. Barrett ordered his crew into one of the ditches while firefighters from the Scotch Creek fire department worked the other side of the road.

Their efforts to protect the bridge appeared to be holding, but farther away Barrett could see that fire was encroaching. He repositioned his crews farther west, trying to protect a long, straight stretch of the Squilax-Anglemont Road near the Scotch Creek Transfer Station. After making it across the bridge, fleeing residents still had to follow that road for roughly a kilometre through densely packed trees that were now entirely aflame. From the transfer station, the road bends slowly towards the south, and then it's a more or less straight shot to a high steel-and-concrete bridge over the South Thompson River, and the Trans-Canada Highway beyond. By the time Barrett reached the transfer station road, fire was burning through the trees on both sides and rolling across the road itself in waves.

Though they were less than a kilometre apart, Barrett and Phillips couldn't reliably communicate with each other. Radio communication across the region was either failing or jammed with traffic from many firefighters trying to relay messages to each other at once. In the chaos, Phillips didn't realize how bad the fire beyond the bridge had gotten. As Phillips and his firefighters continued defending the bridge itself, they waved dozens of cars across, believing they were sending evacuees into safety. "I sent my daughter out that way without a thought," Phillips said. "I thought it was safe."

He didn't know that the rank 5 fire he'd been warned about had already reached the road by the transfer station, where Barrett and his crew were now fighting desperately to hold it back. As the fire along the road intensified, they were able to buy time for about forty vehicles to get out of Scotch Creek before flames made the road impassable. Phillips's daughter and his two toddler grandchildren were likely in one of those final forty vehicles to make it out along the road around 3 p.m., feeling the temperature rise inside the car as flames lashed at them from the roadsides.

Just as the last of those cars made it out, Barrett's crew started running out of water. They were forced to disengage from the fire and head west towards Lee Creek to refill. When they returned, they found themselves in the middle of a firestorm. A thunderous roar filled the air, with ember showers falling everywhere. There was no way to get through and back to the bridge. Barrett looked down at the firehose in his hands. "We were bringing a knife to a gun fight," he said.

As the fire intensified around them, they retreated to Tsútswecw Provincial Park, just west of Lee Creek, and started preparing for the worst-case scenario: a burnover. The park included a wide-open field and large asphalt parking lot—good locations to use as safety zones if they needed them. If they were burned over, Barrett's firefighters could still shelter in their fire trucks and spray water over themselves like a giant water umbrella. It wasn't ideal, but it was something. As they readied their defences, Barrett told his crews to call their loved ones.

By 3:30 p.m. when Jesse's task force reached the sawmill, workers there were already surrounded and the fire in the area was too dangerous to fight. The Bighorn and Firestalker unit crews working nearby were ordered to fall back and shelter within the mill itself, taking Jesse's task force along with them. But they didn't just run, they fell back strategically, leaving pumps, hoses, water bladders and sprinklers set up in key areas where they'd need them again once the fire front passed. It was all part of the layered lines of defence firefighters frequently build when they expect to mount a fighting retreat. Firefighters sheltered at the mill for an hour as the flame front roared around them. Then, as planned, they emerged to re-engage it again.

WHEN FIREFIGHTERS LEFT THE MILL AND ARRIVED BACK AT Holding Road, they found not only a fire, but anger. By now multiple structures were fully ablaze, and many others were threatened. As they set about trying to protect what was left, Jesse's task force and the other BC Wildfire Service firefighters encountered furious locals who felt they'd been abandoned to face the fire front alone. Jesse saw people in shorts and T-shirts, even flip-flops, racing around and trying to extinguish spot fires.

"We tried to explain that they weren't really helping," Jesse said. "We were starting to see trees coming down, burning power poles, downed electrical wires." He and his firefighters were focused on trying to prevent people from getting hurt amid the catastrophe. The community's frantic and uncoordinated efforts risked dividing firefighters' attention. Along with trying to battle the fire itself, Jesse's task force and the Bighorn unit crew were also trying to save townspeople from danger. It was nearly impossible to do both well.

Amid this unfolding pandemonium, frustrated locals started taking the firefighters' pre-positioned pumps and hoses, trying to use them to defend their properties, including removing structure protection equipment from a key bridge three times in one day. Having just watched professional firefighters retreat to the mill's safety zone while the fire front swept through, they believed this equipment had—like them—been "abandoned" by firefighters.

It can—understandably—feel impossible for a homeowner on the verge of losing everything to recognize why their house is being triaged as a lower priority; why critical infrastructure like bridges or fire halls or water treatment plants that no one lives in must survive, even over people's homes. Fighting a wildfire often means hard choices. If the North Shuswap's critical infrastructure was lost, it wouldn't just be weeks or months before residents could return, it would be years. But that can be a hard reality to grasp when hungry flames are reaching towards the home you raised your children in.

Sometime around 4 p.m., it became apparent the fire would reach the four-hundred-person fire camp at the Squilax airfield, and the few firefighters still in the camp prepared to evacuate. Firefighters who were there described scenes of absolute chaos. The Brazilian firefighters, including those who'd survived the entrapment the night before, were positioned with water tankers in a defensive arc around the camp, facing the flames that were now racing through the treeline surrounding them. Members of the incident command's three-person information team were trying to coordinate collecting the belongings of everyone who was deployed fighting the fire. The insides of many firefighters' tents were in disarray—no one had expected when they went to work that morning that they might not have tents to come back to.

There was no time to dismantle the tents themselves, so firefighters focused on shovelling whatever belongings they could into garbage bags, duffels, anything they could find, and packing them into the handful of vehicles still left in camp. These weren't just trinkets or dirty sleeping bags—there were laptops, medication, even the international firefighters' passports. All of it was at risk of burning.

At around 5 p.m., unit crew Bighorn arrived back at the airfield and threw themselves into the demobilization efforts. Embers had started to land directly inside the camp, igniting spot fires that were burning some of the tents. Even the camp's catering staff were trying to douse dumpsters full of cardboard that was now on fire.

While all this was unfolding, the fire was burning ever-closer to Scotch Creek itself. Flames were less than two hundred metres from the Scotch Creek gas station, in the centre of town. And then came more bad news. Jesse's task force received a call ordering them back towards Celista. Structure protection equipment at another bridge east of Scotch Creek had been stolen. Without it, fire threatened the only route east away from the flames. With the road beyond the Scotch Creek bridge now covered in fire, both east and west escape routes were compromised. The only way out of Scotch Creek now was by boat. Dozens of residents made the treacherous trip across Shuswap Lake, lashed by whitecaps and forty-kilometre-per-hour winds. Phillips's second daughter and her children were in one of those boats, while his wife—a firefighter herself—led a convoy of vehicles east over an untested logging road towards Seymour Arm.

At just after 7 p.m., Reynolds blew an air horn at the fire camp, signalling the final retreat. Unit crew Bighorn's supervisor led his firefighters away from camp. He expected to see a tight convoy of vehicles—the formation they were trained to drive in—but instead met more chaos. One person was almost run over. The driver of another truck looked in her rear-view mirror and saw a Costa Rican firefighter running after the vehicle. When she stopped, the man yanked open the truck door and jumped in. In the chaos, his crew hadn't noticed he was missing, and he'd accidentally been left behind.

In Scotch Creek and elsewhere, incident command was now ordering even structural firefighters to leave. Across the region, the fire was now burning at ranks 5 and 6: the most extreme behaviours on the six-point scale, and far beyond what is safe for firefighters to face. "We don't put firefighters in front of rank six fire," Healey would tell me months later, underscoring just how dangerous it was for so many residents to have stayed behind through the firestorm.

Contractors and other structure protection specialists who had been sheltering near the marina were ordered to evacuate around 8:45 p.m. as the last of the evacuation orders went out. They joined a convoy of firetrucks that made the harrowing journey back down the Squilax-Anglemont Road. As they drove, they could see fire already climbing the slopes of Skwlāx Mountain across Shuswap Lake. The fire had made an incredible twenty-kilometre run in less than twelve hours, one of the fastest wildfire runs in BC history.

Despite the risks, many locals still refused to abandon the town. Mark Acton and his father Dean were still racing around fighting spot fires where they could. In the years before the Shuswap firestorm, they'd invested in significant firefighting supplies, including a dump truck with a large water tank, pump and hoses, and a secondary water tanker to help resupply it. Now they put those tools to the test.

Even with the wall of water from the high-volume sprinkler system holding the main fire front at bay, embers still rained down over large swaths of Scotch Creek. Mark and Dean Acton and other volunteers battled the blazes the best they could, knocking down spot fires as they started. At one point, the back of the local Home Depot caught fire. Locals ripped the siding off and doused the flames, likely saving the building.

Another spot fire caught at the construction yard of North American Log Crafters. The property was packed with half-constructed log homes, piles of treated whole logs and all manner of equipment. The Actons and others fought the fire there as it ate into log home assemblies. They used heavy equipment to bulldoze the fire back into itself and smother it with dirt. At one point, several nearby barns were threatened, but local residents managed to beat back the flames with firehoses, likely preventing the ember storms those structures would have caused from igniting other nearby homes.

Five kilometres west, as the convoy of fleeing firetrucks headed towards Highway 1, the firefighters were forced to drive through the flame front, which had by now passed beyond the airfield camp and was racing through the trees towards the last remaining bridge out of the area. One firefighter described it as "inferno road." As they drove south, flames roared in the ditches and tree canopy to their east and west. Ruddy-orange smoke occasionally obscured the road completely, and firefighters feared what would happen if a tree blocked their path, or a vehicle broke down. The firestorm was so intense and evacuation orders so late in coming, one firefighter later told me that as they retreated, they were convinced they would need cadaver dogs to search through the rubble when they returned.

Healey was among the last of the firefighters to leave the camp, ensuring everyone else had already gone. But when they reached the Squilax bridge and Highway 1, he made a startling discovery. The fire had moved so far so fast that the regional district hadn't yet set up any roadblocks. The highway east of the bridge was still open, sending unsuspecting drivers straight towards the flames.

As they reached the highway, the convoy turned west towards Kamloops, but Healey—alone—turned east, driving through the fire's eastern flank, trying to stop traffic and prevent terrified motorists from driving headlong into the wildfire, now burning freely on both sides of the highway. He called the Kamloops Fire Centre and requested the immediate closure of the highway, then stayed in place blocking it himself until the Ministry of Transportation could get a formal traffic control point set up as the fire roared its way up Squilax Mountain above him.

AFTER THE FIRES IN THE NORTH SHUSWAP MERGED, THEY OFficially became known as the Bush Creek wildfire. It continued to burn in the hills above the towns of Sorrento and Blind Bay, on Shuswap Lake's southern shore, for weeks.

The damage caused by the August 17–18 windstorm was staggering. Veteran firefighters told me they'd never seen fire behave like that before. Lee Creek which, according to fire behaviour modelling had faced the biggest risk, miraculously emerged almost unscathed—likely thanks at least in part to the planned ignition. The fire had blown across the eastern section of powerline, where it had caused hours of nail-biting for Ingham and nearly killed the trapped Brazilians, before sweeping into Meadow Creek and beyond. But to the west, where Coast Romeo had also burned off the powerline corridor, the lines had held. The final fire perimeter map (released months after the blaze) showed a clean, straight edge running along the powerline, creating a pocket free of fire above Lee Creek. Everywhere else the fire had burned right to the shoreline.

Scotch Creek lost dozens of homes, its recycling centre and bottle depot. A trailer park had gone up in flames, leaving the disfigured husks of cars melted half into the ground. Small rivers of molten metal—the remains of aluminum hub caps and catalytic converters—meandered across scorched asphalt and gravel before cooling into strange and ugly works of modern art. The town even lost its fire hall.

Further east along the lakeshore, more homes were gone, burned down to their foundations. At one of them, a husk of concrete sat with dark, empty windows like the eyes of an enormous skull staring out over the waves. The property's dock reached out into the water and a fake decorative palm tree was somehow still standing, its plastic fronds only slightly melted from the heat. It was like this for kilometres along the road, house after house burned down to nothing, with yet more destroyed properties lining the hillside behind them. To the west, the Squilax gas station was gone—firefighters had heard it exploding behind them as they fled—and many homes across the Skwlāx te Secwépemcúlecw First Nation reserve were also lost. In all, more than 160 homes and other buildings burned across the region, with 130 of them completely destroyed.

In the days following the firestorm, dozens of locals who'd stayed behind continued fighting smouldering hotspots around their properties. Professional firefighters kept up the battle as well, but often far out of sight—chasing the fire front up into the hills south of Highway 1 in areas far from the view of locals. With the chaos and drama of the windstorm now over, the grinding work of finally reining in the fire itself continued. Crews went back to swinging Pulaskis, cutting fire guard, and dragging hose line across the mountains for miles.

By then, relationships between local residents and provincial authorities were frayed well beyond repair. Police roadblocks began going up the day after the windstorm to keep people out of the evacuation zone, and local residents, speaking by phone to reporters stuck outside, described RCMP officers swarming into the area.

While it was at first the fire that kept many local residents trapped, some soon found themselves stuck on their properties or dodging police patrols when they tried to leave. British Columbia's provincial laws do not allow police to physically remove someone from their own private property inside an evacuation order area, but as soon as a resident sets foot off their land, they are technically breaking the law and subject to arrest. This had the effect of creating a cat-and-mouse game between locals and the police. Those who continued defying the evacuation orders had lost all faith in the wildfire service and local fire departments. There were still plenty of hotspots burning in the forests nearby that they felt needed to be doused. They worried another windstorm might reignite them.

But that meant leaving their own properties to follow the fire where it burned. Social media started filling up with viral videos of exhausted-looking police officers arguing with equally exhausted local residents, urging them to leave the area or—at the very least—return to their homes, and threatening to arrest them if they didn't.

It didn't take long for the evacuation-defying locals to start running short of food and fuel. Community efforts to resupply them were repeatedly stymied by police and government officials, who at one point refused to allow a tractor-trailer full of donated food to enter the area. Residents instead relied on their knowledge of local logging roads to get around some roadblocks and smuggled in supplies with boats, sometimes late into the night to avoid police patrols on the water. Many described the North Shuswap as having turned into a police state.

Compounding this were the simmering tensions about the backburn itself. In the days and weeks following the firestorm, wildfire service officials repeatedly described the backburn as a success, saying it had helped to save many homes, especially in Lee Creek, a position it maintains today. But it seemed to locals that they were not getting the full picture from the outset. Everyone, it seemed, knew someone who knew someone in the fire service whose accounts of what happened didn't match the statements of public officials. At least some of this was due to the fire service's own internal struggles to figure out exactly what had happened during the windstorm's chaos.

The situation in the North Shuswap worsened as the weeks rolled on. The fire season was far from over, and there were still huge areas of the North Shuswap that needed attention from the overstretched provincial fire services. The more that locals voiced their frustration and continued defying evacuation orders, the tighter the police and government clamped down, at one point deploying spike belts at roadblocks to prevent frustrated residents from simply driving around them. The situation festered until

August 24 when a self-described "convoy" of angry residents from surrounding communities—many spouting well-worn conspiracy theories—challenged one of the RCMP's highway roadblocks. Inside the evacuation zone, wildfire crews reported at times having garbage thrown at them by angry local residents.

It wasn't just firefighters whom residents were upset with. Though many people channelled their anger in constructive directions, seeking out and dousing stubborn hotspots, or helping to feed their exhausted neighbours, some took out their frustration on the handful of journalists who managed to evade police roadblocks and get inside the evacuation zone. Several residents threatened a photojournalist, and others accused reporters of being mouthpieces for a government they'd begun to see as complicit in the destruction across their communities. The resentments ignited by the fire and fanned by the windstorm would smoulder for years.

I FIRST MET DAVE DYCK IN APRIL 2025, ALMOST TWO YEARS AFter the fire. Raised as a Mennonite in Ontario (with stints in Mexico and Brazil), Dyck has the rough-skinned handshake of someone who works with his hands. He invests his trust not in government but in community. "Helping each other—it's what we were raised in," he said. Though almost two years had passed since the fire, the impacts are still obvious in the burned stalks of trees encircling his property. He bought his first Meadow Creek place, he said, because he wanted to get back to his farming roots. He's since added others nearby to his portfolio, building houses by hand and flipping them for profit.

Sadie, a cat with silky black fur, stalks his house, always on the prowl for cuddles. The kitchen door is hung with an impossibly heavy wreath of garlic, its scent mixing with the aroma from the pile of sticky pot buds in a bowl on the kitchen table. Jazz piano plays from a stereo in the open-concept living room. Dyck's garage is filled with the kind of high horsepower tools and toys—a side-by-side, a four-wheeler, trucks, tractors, a BobCat—common among folks who live at the end of dirt roads hours away from the nearest Starbucks.

Dyck has an easy, slightly mischievous smile above a blond soul patch, and long, wild hair he keeps tied in a ponytail behind his head. He speaks softly but a lot, with a gentle cadence that quickens only for two things: when proselytizing about the health benefits of hyperbaric chambers (he runs a clinic in town with one and claims it can cure all manner of ailments) and when he's talking about the wildfire.

The impact the fire had on him came through the more he spoke. When I arrived at his Meadow Creek home, he was busy tidying up his properties, preparing to sell them. He greeted me with a warm smile but no other pleasantries. Instead, he dove right in.

"This was arson," were the first words he said. "This was intentional. How could it not be?" he said, sweeping his arm in a wide arc towards the platoons of blackened tree trunks surrounding his property. He told me in words laden with pain and anger about how he had watched the aerial ignition end, and how he claimed to have seen four distinct mushroom clouds explode into the sky. He mentioned directed energy weapons (a well-travelled and widely debunked conspiracy theory) more than once. He was convinced that this was done by the government on purpose, though the alleged motives for something so sinister were never clear to me.

As Dyck spoke, he wound himself tighter and tighter into a ball of angry anxiety, and as I listened, it was easy to understand why. "You're going to tell me that I'm not allowed to fight the fire on my own land, and when I try, you're going to put out fucking spike belts and roadblocks and try to arrest me? Fuck you!" he shouted at no one in particular, then paused to take a deep breath. "Sorry, I still get pretty angry about all this."

Despite the trauma still obvious in his voice, Dyck told me that watching the community come together to protect itself and each other during the chaos of August 17 and 18 left a deep impression on him. It's what he grew up seeing and believing to be most important. "It was all 'Hey, what do you need? I've got a tidy tank, and we need it over there. Okay, it's on its way,'" or "Hey, can you give me an apple, a sandwich? I haven't eaten all day."

Dyck said it was eight days before a single person in a red BC Wildfire Service shirt showed up to Meadow Creek to help fight the fire, though this is contradicted by internal BC Wildfire Service records. Healey's logbook (which I obtained through a Freedom of Information request) and Phillips's recounting to me both made it clear that structure protection specialists did work in Meadow Creek the night of the backburn, though they didn't arrive until hours after the fire started burning near Dyck's property, and it's possible he didn't see them. It's also possible he wouldn't have recognized them as BC Wildfire Service crews if he did. Structure protection specialists are often brought in from other municipal services or hired as contractors. They don't always wear the red shirts and blue pants of the BC Wildfire Service, but whatever Nomex uniform their own fire department or contracting company has issued them.

Dyck said he and his neighbours did their best for more than a week with what felt like no support, fearing that at any moment a wind shift or a nasty spot fire could wipe out their entire life savings and leave them destitute or dead. And when Dyck finally did encounter some wildland firefighters, he said they were so young and green that they looked terrified. "They were teenagers. They were shaking in their boots," he told me.

Meanwhile, he characterized himself and his neighbours who stayed to fight the fire directly, as well as those who helped them, as a guerrilla firefighting movement. He stressed to me that they were treated like criminals, forced to evade police, dodge roadblocks, drive through ditches around spike belts and smuggle in supplies like food and fuel, all to keep the movement going.

Much of this effort was organized around Karl Bischoff's farm in the hills above Celista. A former wildland firefighter himself (as well as a champion lumberjack), Bischoff is a squat, rough-speaking man who doesn't waste time, or words. His farm, which has stood since the late 1800s, has long been a rallying point for locals in the region whenever a crisis strikes. For days following the August 17–18 firestorm, Bischoff's farm became a de facto command post for dozens of volunteers. Twelve pickup trucks were pressed into service as initial attack vehicles with thousand-litre water tanks strapped into their truck beds. Bischoff calls them redneck firetrucks.

After the firestorm itself, which Bischoff said sounded like a jet engine even from miles away, more volunteers kept coming. People brought RVs and camped in his yard, with people cooking in shifts to feed everyone. Most of the food had to be smuggled through police roadblocks when friendly officers were on duty, willing to turn a blind eye.

They worked "like bandits" Bischoff said, chasing smoke up into the hillsides around their homes wherever they saw it. "Hey, there's fire here," one guy would say. "Okay, send a couple guys." "Okay, now it's over here, send some more." It went on like that for days.

Bischoff himself put over five hundred kilometres on his quad in ten days, driving around the area scoping out hotspots, assigning crews, and sometimes extinguishing the fires himself. Friends of his drove even farther.

On the eighth or ninth day (in the chaos since the fire, Bischoff forgot exactly which), a crew of BC Wildfire Service firefighters arrived and triggered what Bischoff described as a "Mexican standoff."

The firefighters were frustrated with the number of local residents racing around in the fire zone, and with the lack of communication between them and the professional crews. Incident command structures on a wildfire typically require the careful coordination of resources, and rogue firefighters moving around in places that the higher-ups couldn't keep track of could compromise firefighting. Water bombers can't drop on an area if there might be unknown people below them. Crews can't conduct ignitions if there could be locals on a mountainside somewhere above them. But that mattered little to Bischoff and the other volunteers. All they could see was a firefighting job that needed doing, and few professional firefighters around to do it. Bischoff argued with the BC firefighters. "You guys created a mess, and we're trying to clean it up," he said.

The tension started to worm its way between professional fire crews as well. Twenty kilometres southwest, across Shuswap Lake in the hills above Sorrento, the fire continued to burn, and crews continued to fight it. One night in late August, Shuswap Fire Chief Ty Barrett's crews got called out again. The fire was threatening a chicken farm, burning through the forest around it and heading into the fields. When Barrett arrived, he found a wildland fire crew preparing to burn off the field between the fire and the barn. Barrett didn't understand why they weren't attacking the fire with hoses and water like structural firefighters would, and the two crews argued over how to proceed.

The fire was finally declared out on December 13, six months and one day after it started. It had burned nearly forty-six thousand hectares, destroyed almost 200 homes and forced more than eight thousand residents to flee. But even with the fire itself finally out, anger in the community continued to smoulder below the surface. Local residents mounted a campaign to seek accountability from the government for the planned ignition, securing hundreds of signatures on a petition calling for an investigation. The Forest Practices Board launched one, though three years later it still had not released any findings. The longer residents went without satisfying answers, the more their campaign took on a narrative that the fire service was either negligent or nefarious, or both.

DAYS AFTER THE NORTH SHUSWAP FIRESTORM, I GOT MESSAGES from firefighters who were upset with how the planned ignition and its aftermath had been handled. Some claimed the wildfire service was trying to cover it up. This wasn't entirely true—though the service did fail to notify its employees' union about the Brazilians' entrapment and at first seemed resistant to a joint investigation with the union. The firefighters' union reps were so infuriated by this they took the extreme step of sending a letter to the Minister of Forests demanding that IMT 5 be stood down until a formal investigation could be completed, something that—in the middle of wildfire season—the ministry refused to do.

The more I looked into what had happened, the more I realized that what the public had been told about the ignition during the disaster was not the complete truth. Based on leaked internal documents, I wrote a newspaper story in December 2023—months after the fire itself—about the Brazilians' entrapment and the other firefighters' narrow escape. This work took months to accomplish and showed for the first time that there were significant details about what had happened that the government was not revealing.

From the outset of the Adams Lake fire, many people in the North Shuswap communities felt unheard, ignored and talked down to. Their worries about the fire in its early days felt brushed aside, and the sequence of events that followed only seemed to entrench those feelings. Understanding that yours is only one of hundreds of communities threatened by fire, or that larger population centres must by necessity take priority when resources are stretched thin—those are difficult things to expect people living through a traumatic experience to grapple with.

While I don't believe Dyck's incredible assertion that the BC Wildfire service allowed their communities to burn on purpose, given everything he and Bischoff and their neighbours went through, I can understand why they feel the way they do. It would be easy to play Monday morning burn boss, to blame what happened in the North Shuswap on inexperienced rookie firefighters, or an overconfident incident commander, or—as many have—a fire service they think doesn't care about them. In the aftermath of a disaster, it's tempting to go looking for a bad guy. Reality is often far more complex and its answers are less satisfying.

What happened in the North Shuswap was the result of long-standing systemic problems colliding with the new reality of modern wildfires' terrifying destructive power. The forests around homes in the North Shuswap were loaded with fuel. Years of drought had parched the land. A lightning strike gave the fire the high ground from the start, during a summer in which the whole country seemed to be on fire. The firefighting system was strained to its breaking point, firefighters both on the ground and in command forced to choose between multiple bad options, making coin-flip calls in lose-lose situations without enough resources to do their jobs properly.

The government's refusal to be straight with residents about exactly what had happened and why in the days, weeks and months after the fire shattered the communities' trust. Once you've lost the public's trust, no amount of carefully crafted bureaucratese or clever message management is going to win it back.

Along with battling wildfires themselves, Canadian firefighting agencies too often find themselves stuck in a battle for the narrative of the incident as well. This is misguided and wrong, an outgrowth of Canadian bureaucracy's broader instincts towards secrecy and strategic public relations over transparency and accountability. And these instincts run deep. At a wildfire journalism conference I attended in June 2025, a senior wildfire official said, "Transparency is a long game. It's about mutual respect, about seeing that we're working towards the same outcome: a public that is safer, informed, and ready." It's difficult to argue with the importance of mutual respect. And every serious journalist I know has deep respect for firefighters and emergency workers. We are all working towards a safer, more informed society. Public service is, after all, at the heart of what drives most journalists in the first place. (Believe me, I could enjoy a far cushier lifestyle if I worked in public relations instead.)

At the same time, we're often accused of "gotcha journalism" whenever the government doesn't like the framing of a particular story. When I pull out my cameras, jokes about Peter Parker abound. What the senior wildfire official seemed to miss, what fire agencies in Canada so often appear allergic to, is that journalists also serve another critical function: it's our job to hold them to account. The relationship can't always be collaborative. Suggesting that journalists must earn the trust of government before the government can or will be transparent left me floored.

I'll be the first to admit that Canadian journalists need to take accountability for wildfire coverage that is sometimes shallow, sensational, or missing important context. We can, and should, do better. But those mistakes are not a license for public agencies to simply throw up roadblocks (figurative and literal) because they don't like what we write. Transparency from government is a requirement for public trust, not the other way around. In a democracy, that's not negotiable. It's also how we get better. Fire agencies, including the BC Wildfire Service, frequently complain that the public doesn't understand them or the pressures they face. They bemoan news coverage that misses the point or misconstrues tactics like planned ignitions or indirect attack. But keeping us stuck outside roadblocks many kilometers away from what is happening undermines public safety in the long run. It ensures journalists won't fully understand what they are writing about, and ultimately weakens our reporting, compounding the problem. Withholding critical public documents for years and hiding uncomfortable truths behind lines of blacked out text damages the very trust they claim is necessary.

It's also how we get better. Fire agencies, including the BC Wildfire Service, frequently complain that the public doesn't understand them or the pressures they face. They bemoan news coverage that misses the point or misconstrues tactics like planned ignitions or indirect attack. But keeping us stuck outside roadblocks many kilometers away from what is happening undermines public safety in the long run. It ensures journalists won't fully understand what they are writing about, and ultimately weakens our reporting, compounding the problem.

At Scotch Creek, during both the fire and its aftermath, the BC Wildfire Service lost the battle for the narrative thanks in large part to public messaging that local residents simply didn't—or wouldn't—believe. The impact of that has undermined the North Shuswap public's trust in the fire service for years. It was a pattern I'd seen play out in rural communities before, especially during a crisis when larger centres get squadrons of water bombers as the skies fill with smoke.

But what happens when communication fails in larger cities? Is mistrust of the fire service simply a quirk of rural life in Canada, or was there something more at play? In the North Shuswap, communication failures weren't only responsible for breeding conspiracy thinking, they nearly got people killed. What happens when those mistakes affect not just thousands of people, but tens of thousands? I didn't have to look far for answers. The same weekend Scotch Creek burned, a huge wildfire was bearing down on Yellowknife, the capital of the Northwest Territories.

CHAPTER 5

RADIO SILENCE

The city of Yellowknife sits perched on the shores of Great Slave Lake, a northern inland sea surrounded by rocky Canadian shield in the upper reaches of Canada's vast boreal forest, some of the most fire-prone landscapes in the country. Thick stands of black spruce and tamarack slowly give way to balsam fir and dwarf spruce before finally conceding to the barren Arctic tundra entirely. It's nearly one thousand kilometres north of Edmonton, Alberta, as the raven flies. In the 1800s, settlers gave the name Yellowknife to a nearby river after the copper knives carried by its original inhabitants, the Wiiliideh Yellowknives Dene who hunted, trapped and fished along the river and lakeshores for centuries. To them it has always been known as Somb'a K'e.

Since its founding, Yellowknife has been a hub for mining. Its first significant development was a collection of homes and businesses on a peninsula jutting out into the lake, but with the discovery of gold in the 1930s, it quickly outgrew the rocky outcropping, and a larger centre was built on the main shore. Between the 1940s and the early 2000s, Yellowknife hosted five major gold mines, the largest of them the now-infamous Giant Mine, with underground shafts and tunnels snaking more than fifty-six kilometres underneath Great Slave Lake's rocky shoreline.

Today Yellowknife is home to more than twenty thousand hardy souls who brave its -30 degrees Celsius winters with relative ease, weather so cold it covers the lake in several feet of ice thick enough to drive firetrucks to the doorsteps of houseboaters who live year-round offshore. I spent a winter living in a tiny cabin on the lakeshore in Old Town in 2011, crawling under the floorboards to chip away the ice that formed when frigid winds blew off the lake and froze my shower drain pipes. I pretended cabin life made me tough, but in reality I jumped at the chance to move into a house with proper plumbing and a roommate in town that spring.

Working at the *Yellowknifer* was my first real newspaper job, as the sports reporter/photographer/editor for the Northern News Service's chain of papers. I remember the first time I saw the northern lights—really saw them—while walking back over the frozen lake from a late-night jam session with friends at one of those houseboats. By the end of the winter, the lights had become so routine they hardly merited an acknowledgement when I'd glimpse them from my bedroom window.

While its winters are rough, summer in Yellowknife is beautiful, and warm enough to bask in shorts-and-T-shirts sunshine at the city's famous Folk on the Rocks music festival. Like most industry towns, Yellowknife is a city that knows how to have fun. Tourists and locals alike often emerge bleary-eyed from the Gold Range pub to find daylight at 2 a.m., exhausted from a night spent dancing to the raucous house band Welders Daughter. Southerners frequently confuse Yellowknife with Whitehorse, the capital of the Yukon, but the two are really quite different. Whitehorse is like the Vancouver of the North; you can live out your Jack

London frontier fantasy without having to give up your pumpkin spiced latte. Yellowknife is more like Edmonton, a somewhat gritty government town with a bootstrapping industry side that still manages to have art galleries and decent coffee.

As the capital of the territory and its fortyfour thousand people, Yellowknife is also a town where almost everyone knows someone who works for either the city or the territorial government.

Yellowknife is easily one of the most isolated cities in Canada. There is only one two-lane road in or out. It's 100 kilometres to the tiny hamlet of Behchokǫ̀, 315 kilometres to Fort Providence, and 480 kilometres to Hay River. Even then you still haven't left the territory. In fine weather on empty roads, it's a fifteen-hour odyssey by car to reach Edmonton, the nearest city of any significant size.

Because of this isolation, everyone tends to know everyone. There is something about living in the North, a feeling of being on the periphery of things that binds people together quickly. Like most northern capitals, Yellowknife is a place where gossip spreads like wildfire.

During the summer of 2023, there were more than three hundred fires across the southern portion of the territory, including a nearly seven-hundred-thousand-hectare monster straddling the intersection of the BC, Alberta and NWT borders, and an eighty-thousand-hectare fire west of Enterprise that threatened to cut off the highway access south into Alberta.

Wildfire ZF015 was discovered on June 28, burning in dense black spruce between Yellowknife and Behchokǫ̀. In late July, it made several big runs west, first destroying four homes in the tiny community of Rae, then levelling fifteen homes in Behchokǫ̀. Firefighters were brought in from Alaska and elsewhere to fight it, but efforts proved challenging, in part because of the area's remoteness and lack of forest service roads. It was extremely difficult for firefighters to attack it directly. It also wasn't the only worrisome fire nearby. Another blaze was burning roughly forty kilometres north of Yellowknife, and a third ignited August 2, forty kilometres south of Dettah, a mostly Indigenous community on the eastern shore of Great Slave Lake. By early August, the trio of fires had Yellowknife more or less surrounded.

When all three fires started menacing the Yellowknife region, firefighters were already on the back foot. With only around 140 Type 1 firefighters in the whole territory, and 250 Type 3s, NWT Fire did not have enough firefighters to handle the flames on their own, and would typically rely instead on outside resources being flown in to help. In a normal season, that can work, but amid a fire season as extreme as 2023, available resources were few and far between. Complicating matters more, unlike in BC, the NWT didn't have any highly trained incident management teams capable of managing large, complex fires. As the fires grew in size and complexity, firefighting leaders struggled to maintain operational control. Firefighters from different municipalities and jurisdictions sometimes wound up working at odds with each other, executing different firefighting missions without enough coordination.

Organization outside the firefighting apparatus itself quickly fell to shambles. Elected officials did not understand how incident command structures worked or what they could reasonably expect from firefighters. "Political interference was more cumbersome than operations, yet no one wanted to make decisions," one senior government leader later wrote.

As the fires got bigger and more dangerous, the NWT's firefighting agency struggled to keep track of how big and how dangerous, and often couldn't reliably predict the fires' movements. Monitoring the fires was a challenge. Many outlying communities felt they were put unnecessarily at risk. Students were hired to track the fires and monitor fire behaviour, but that work was inconsistent at best. Critical fire monitoring data wasn't used in a timely or coordinated way. Making things worse, the speed of the fires often overwhelmed the territory's ability to keep up. Along with struggling to track many of the fires' actual movements, the territory also lacked sophisticated wildfire modelling tools capable of generating the kind of spread predictions that fire behaviour analysts and incident commanders rely on to decide everything from in-the-moment tactical deployments to issuing evacuation alerts and orders. What little predictive modelling NWT firefighters did have was sometimes either misunderstood or disregarded.

In one case, after evacuating because of a fire, members of the Kátł'odeeche First Nation were told, based on fire monitoring by the territorial government, that it was safe to return to their community only for the fire to come back that night, forcing another evacuation.

On August 13, there was another alarming close call, one that would leave officials and experts rattled, and that contributed to the eventual call to empty Yellowknife itself.

On August 12, driven by ninety-kilometre-per-hour winds, wildfire SS052, which was burning more than two hundred kilometres southwest of Yellowknife, made an explosive run towards the predominantly Indigenous community of Kakisa, ripping through more than forty kilometres of forest in a single day, spreading at more than double the speed fire behaviour analysts thought possible.

The next day, fire officials called a virtual meeting with elected officials to update NWT communities on the situation and which communities were most at risk. But somehow representatives from the community of Enterprise, four hundred kilometres south of Yellowknife, were not invited—even though that community was facing extreme risk. It was the chief of the nearby Kátł'odeeche First Nation who noticed Enterprise was missing on the call and reached out directly. As soon as it was clear how close the fire was, Enterprise ordered an immediate evacuation at 2:30 that afternoon, with its roughly one hundred residents racing to escape by 8:30 p.m. The fire roared into town just after 9:30 p.m. and erased 90% of the town.

As the situation worsened across the territory, the emergency communications system began to crumble. In some communities, residents didn't receive evacuation alerts to their phones even after signing up repeatedly for the territory's automated digital alert system. When evacuation orders came for places like Behchokǫ̀, about an hour outside Yellowknife, some people were prepared and evacuated quickly, while others did not. Many believed they had more time than they did, because the last they'd heard from officials the fire was still far away. In some cases, residents wound up fleeing through flames, driving directly through the fire's front as it swept into towns like Hay River.

Ollie Williams had watched and chronicled these breakdowns for weeks. Williams is a gregarious Brit with a penchant for understated humour that belies a quiet determination—a quality he tends to downplay by poking fun at himself. Williams helped found Cabin Radio in 2017 with a handful of other locals after giving up a prestigious job with the BBC's sports division in the UK and moving to Yellowknife, following a woman he met while covering the 2010 Olympics in Vancouver. When the Canadian Radio-television and Telecommunications Commission refused to give Cabin Radio a broadcast license, Williams and his co-founders were undaunted and made it an internet station instead, broadcasting news and interviews across the territory from their unassuming headquarters on Yellowknife's Franklin Avenue. Within only a few years, they'd developed a fiercely loyal audience, attracting listeners with a brand and sound that was warm, friendly, and adventurous. When he's not in the studio, he spends his time with his partner, Liny Lamberink, herself a reporter at the local CBC station, their cat, and a dog who loves to greet visitors when the front doorbell rings.

Cabin Radio's reporting during 2023 wildfire season, as community after community got hit, earned them a reputation for speed and reliability. Listeners and readers started bookmarking the website, returning over and over for the latest on the crisis, trusting it more than the government's official channels. Cabin Radio's coverage was so thorough, it formed the spine of Yellowknife's public record of the fire. But it was exhausting work. Williams remembers coming home one night after a particularly long shift to find a pot full of homemade stew waiting on his front porch. A listener had dropped it off, along with a note: "Thanks for what you guys are doing, here's a pot full of dinner." Williams was overcome. He and the Cabin Radio crew had been going flat out for weeks, and he couldn't even remember the last time he'd stopped to eat something. He sat on his front porch and ate the stew amid the growing smoke before going back inside, and back to work.

As fires around the territory continued to worsen, the Cabin Radio crew tracked all the developments in hundreds of stories and live feeds. They told readers how Yellowknifers were banding together to rescue and look after evacuees' pets when two thousand Behchokǫ̀ residents had to flee. They covered the first-ever NWT Culinary Festival when organizers pivoted and started cooking meals for displaced residents, and they shared reader photos and video of homes destroyed only hours after it happened. The outlet became a one-stop shop for all things wildfire related, in contrast to the overwhelmed territorial government, which by mid-July was struggling just to publish timely notices to its social media channels.

As the wildfire situation deteriorated across the province, Yellowknifers began worrying what might happen to them if the fires reached their city.

On July 25, Williams published a detailed interview with Yellowknife Mayor Rebecca Alty, outlining what she described as the plan if the city came under serious threat from wildfires. Williams highlighted how the city's official plan, available online, mentioned "forest fire" only once in seventeen pages to describe evacuees from outlying communities fleeing *to Yellowknife* but included almost nothing about where Yellowknife residents would go if they ever had to evacuate. He contrasted that with the town of Hay River's emergency plans, which ran to one hundred pages and included detailed breakdowns of the conditions that would trigger wildfire evacuations, and how they would unfold.

During the interview, Alty insisted the City of Yellowknife did have its own emergency evacuation plans. They'd been drafted over the past few months, and they just weren't public yet. She agreed that any full evacuation of the city would be an extreme measure of last resort and appeared to downplay the risks Yellowknife faced. Her best advice to residents at the time was not that they should prepare to evacuate just in case, it was to not have campfires or drive ATVs and dirt bikes in the bush. She sounded to me like a politician who did not appreciate the risks her city was facing.

Just days later, she appeared on national television urging residents to leave. She described potentially having to cover the town's key infrastructure with fire-retardant gel in order to save it. "The highway remains open," she told CBC's Renee Filippone on the morning of August 17. "All residents who can, please drive out as quick as you can."

On August 13, the fire threatening Yellowknife started a significant eastern push, breaching control lines and reaching what looked like burning fingers towards the city on either side of Highway 3, the only escape route. It also severed the only fibre-optic internet connection south to Alberta, crashing government systems and hampering its ability to communicate even further. The fire built intensity as it went and flooded Yellowknife with smoke. Despite this troubling development, city officials released a particularly jarring statement on social media that claimed, "Yellowknife is *not* under threat of wildfire.". The word "not" was highlighted in bright yellow text, overlayed on an idyllic scene of float planes tied up at a dock in Old Town and set against a clear blue sky. Despite smoke blotting out the sun that very afternoon, city officials were still insisting that everything was fine.

LIKE MOST GOVERNMENT PROCESSES IN A FEDERALIZED COUNTRY like Canada, the exact structure of evacuation alerts and orders differs slightly between the various provinces and territories. BC and Alberta use a two-tiered system, placing communities on evacuation alert once it's determined there is a chance they might be ordered to flee. This is usually accompanied by information detailing where the fire is and what's being done to fight it at the time the alert is issued, as well as geographic maps outlining the area under alert. The alerts, which are posted to numerous social media accounts, distributed via a downloadable alert system app and sent as releases to local media, typically include links to additional sources of information such as evacuation centres, what residents should prepare to take with them, and how quickly they should be prepared to leave if an evacuation order is given.

The system is imperfect, as the North Shuswap firestorm unequivocally demonstrated. When fires grow explosively and unpredictably, emergency managers are often caught off guard without time to go through the formal alert-to-order process. Sometimes firefighters like Roy Phillips are forced to carry out what they call "tactical evacuations." This means going door-to-door in person, telling residents to flee, even if no formal order has been given. As fire behaviour has gotten more extreme across Canada's West in recent years, these hectic tactical evacuations have become more and more frequent.

In BC during fire season, the province now recommends that everyone who lives in the wildland-urban interface have go-bags with their bare essentials like spare clothes and vital documents packed and ready all summer, just in case. To the great annoyance of my wife, I have my own slightly different go-bag of fire gear including my Nomex, fire boots, helmet, gloves and a respirator that lives in the back of my Subaru from May to October, along with a sleeping bag, a case of bottled water, a camp stove, some freeze-dried meals and enough granola bars to run several marathons.

If there is a gold standard of evacuation communications, it is probably Australia's. Like their volunteer bushfire brigade system, Australia's emergency alert system has evolved by necessity and been shaped by disaster. Until 2009, the system used a three-tier wildfire danger rating system similar to Canada's that consisted of "moderate," "high" and "extreme." After the horrifying Black Saturday bushfires in 2009, which destroyed two thousand homes across the state of Victoria and killed 173 people, the country revamped its system. Today it includes a fourth tier of fire danger: "catastrophic." This rating system is married with a three-tier alerting system that includes direct instructions for what residents should do at each progression. At tier one—the Advice stage—the public is told via alert app messages, social media, emails and news outlets that there's a fire nearby and to stay tuned for additional information. At this point, it's clear that alerts might be ordered, and residents should start preparing just in case. At tier two—called Watch and Act—residents are urged to consider self-evacuating early if they can. Public messaging sometimes frames this in reassuring tones, suggesting for example that, if you are able, why not take the kids and head to a beach or visit family for a few days outside the danger zone. It's explicit there's no need to panic, but best to take precautions if you can.

This is also where the system diverges from Canada's. Along with recommending early, precautionary evacuations, the system recognizes that some people will want to stay and defend their homes and that—with the right equipment and know-how—some people safely can. At this second "Watch and Act" stage, residents are told if they're going to mount a defence of their homes, they need a plan and now is the time to put that plan into action. This is coupled with robust public education campaigns that teach people what it truly means to face down a wildfire on their own. At a minimum, you need defensible space around your home, a reliable independent water source and independent power, like a stand-alone generator, because public utilities should be expected to fail. It's plain that choosing to stay and defend means being able to hold out alone, and that emergency help may not arrive for hours or days. By allowing and supporting this option, the government is acknowledging the differences between an urbanite with no clue how to run a two-stroke fire pump engine and a rancher who may have fought small bushfires her entire life.

The third tier is called Emergency Warning, and it is the direst. It's where advice becomes orders, and those orders include specific deadlines for when and how to leave. It also includes something that Canadian evacuation orders do not contemplate: If fire behaviour is bad enough, it may well be too late to leave and safer to shelter in place.

During the 2009 Black Saturday fires, nearly two dozen people were burned alive in their cars or on the road while trying to flee through flames at the last possible minute. More than 100 others died trapped in their houses. This is the nightmare scenario Australia's system now seeks to avoid at nearly all cost. Residents know from the start where the danger lies, what's being done to manage it, and what they need to do to prepare themselves. Alerts include specific instructions for how best to stay safe as things deteriorate.

One of the most surprising outcomes of Australia's system is that by encouraging people to leave well before an evacuation order issued but still arming them with the knowledge of what it takes to truly stay and defend, Australian fire agencies saw evacuation compliance actually go up. With an informed sense of the danger and stakes, more people were choosing to leave early, avoiding potentially lethal traffic jams.

The Northwest Territories has—at least on paper—a more robust system than BC or Alberta and one closer to Australia's. It includes a third lower tier: a watch-and-listen phase, where locals are supposed to be told when a fire is nearby, what's being done to address it, and that future alerts or orders might be issued. It's meant to allow for a slower scale-up, giving time for remote communities to begin preparing ahead of time.

This system had failed repeatedly all summer, with residents in rural communities often not receiving alerts and updates. Of the few alerts the government did manage to issue, most did not include even basic information about where the fire was at the time, the geographic area covered by the alert, what residents should pack or where such information could be found. Most didn't even include specific danger areas that residents should avoid. They were, in effect, mostly useless.

As the fires burned closer to the capital, the territorial government and the city of Yellowknife couldn't seem to get on the same page. The two governments released information that was often contradictory. The territorial officials shared information that didn't match other sources and were often combative when people asked for more details. Public trust began to crater. By the time the fire breached containment lines west of the city on August 13 and made its push towards Yellowknife's outskirts, the government had largely abandoned its alerting system altogether.

ON THE MORNING OF AUGUST 15, ZOE SHARE WAS TORN, TRYing to decide whether to stay or flee. At the time she was the deputy director of the Yellowknife Women's Society, a not-for-profit that helps support many of Yellowknife's most vulnerable citizens. Among its many programs, the YWS runs a drop in day shelter, a sobering centre, the Women's Centre shelter and a permanent supportive housing space called Spruce Bough. They also do significant outreach to the city's unhoused population, including running a managed alcohol program for people dealing with alcohol addiction. It works by providing small, set amounts of alcohol to clients in an on-site facility with a prescribed schedule. It helps them avoid the crushing—potentially fatal—symptoms of alcohol withdrawal, without spiralling into dangerous binge drinking. As with most social work, stability and consistency are key. Marginalized people who become used to systems that appear not to care about them may disappear when disruption breaks their routines, preferring to be the abandoner rather than the abandoned.

Share had been working at YWS for nearly three years, first as a front-line worker before transitioning into management. Relationships she'd built with clients over years would soon prove vital to maintaining the trust of clients who'd become accustomed to government systems treating them more like numbers than people.

As the fires had crept closer, Share and her colleagues began to realize how difficult it would be if they had to evacuate their many vulnerable, sometimes itinerant clients. But even as the city filled with smoke, the idea they might all have to actually flee was difficult to comprehend. Share felt what many in Yellowknife were grappling with: a sense that a full-scale evacuation was just too big, that there was no way it would really happen . . . right?

That afternoon, as gossip about the fire's eastward run started to spread, Share and her staff started asking themselves whether they had a plan, whether their government partners—or anyone for that matter—had a plan, just to be safe. There were a lot of unofficial sources claiming all sorts of things, from an immanent evacuation to no evacuation at all. Share's social circle was full of people claiming they knew an evacuation was coming, but Share said it was sort of like "Well, you said that last week."

Some government staff were making the same calculations Share was and chose to leave early with their family members, to get them out of the choking smoke. But emails obtained by Cabin Radio reporter Emily Blake show just how inflexible the territorial government was being. At one point, an assistant deputy minister from the Ministry of Environment asked whether their staff could work remotely if they chose to self-evacuate ahead of the fire. Twelve hours before ordering the whole city to empty, human resources responded that anyone who chose to leave before a formal evacuation was in place would not be allowed to work remotely. They would be forced to take vacation or lieu time.

Like many people, Share's partner decided it was time to leave despite no formal evacuation order and started making plans to get out. But Share was caught between her own desire to leave before things got crazy and her sense of duty to her organizations' clients.

And with both city and territorial governments continuing to insist the fire wasn't a threat, Share decided to sit tight while her partner drove out with their dogs the morning of August 15, in part because rumours were spreading that pets wouldn't be allowed if evacuation flights became necessary. Meanwhile Share was trying to coordinate possible charters to fly her organizations' clients out in case they needed them. Her partner said he'd only agree to leave ahead of time with the dogs if Share promised that, when the time came, she would also get on one of the flights with her clients.

Many of Share's friends were government workers, and they were making the same internal calculations she was, though often without the support of their employer. As it became more and more obvious to Yellowknifers that some sort of evacuation, whether partial or full scale, was coming, many government workers had no idea whether they would be considered "essential" and required to stay or be ordered to leave with everyone else.

Other friends of Share's were healthcare workers, and she said many were worried because their contracts included language about not abandoning their posts. Share felt a self-imposed pressure to stay and refused to consider leaving her clients behind. YWS had nearly one hundred people that relied on it one way or another. Shutting down those programs with little notice could have dire consequences for its clients, especially those on the managed alcohol program. The idea of clients going into alcohol withdrawal in the middle of an already traumatic evacuation was something Share couldn't bear.

ON AUGUST 15, CLAIRE BROOKES ROCKED HER FIVE-DAY-OLD son Sterling as she listened to another press conference on TV, refreshing the Cabin Radio website at the same time. Outside, smoke was boiling across the Frame Lake neighbourhood where she lives with Sterling, her husband Stephen Purcell their two Bernedoodles, Murdock and Copper, and two tabby cats named Riley and Merrick. Homes in Frame Lake are mostly a mix of duplexes and single-family bungalows, some set side by side, and interspersed with black spruce, Jack pine, aspen and poplar. The nearby lake is surrounded by rocky ridges dotted with more spruce and pine. As she listened to territorial officials insisting the city was safe, Brookes couldn't help but imagine what it might look like if they were wrong and the neighbourhood around her burned.

That summer the TV news had been full of video showing wildfire destruction across the country and as far away as the Hawaiian city of Lāhainā, which had burned almost to the ground just a week before in an explosive wildfire that destroyed 2,200 structures and killed more than one hundred people. In the past few days, those stories had been replaced by news of fires spreading across Canada, including the one burning less than twenty kilometres away from where she sat.

Brookes had a growing sense of frustration at how little information was being communicated from the city and territorial governments. She was nervous. The smoke kept getting worse, and she was worried it would damage Sterling's infant lungs. Like many people in Yellowknife, Brookes and Purcell agonized over whether to stay or leave. TV news reports kept showing the fire creeping closer and closer, moving in from multiple directions. At one point, she and Purcell considered booking a flight out, but they also had their pets to consider. They refused to abandon the animals but also didn't want to dump them on friends or family and make those potential evacuations even harder.

Williams was watching that press conference too, shaking his head at officials' repeated insistence that they would not be emptying the entire city under virtually any circumstances.

Williams had not expected to become an expert in wildfires, but in the months before the Northwest Territories fires broke out, he had devoured John Vaillant's best-selling book *Fire Weather*, about the 2016 Horse River fire that destroyed whole neighbourhoods of Fort McMurray, the capital of Canada's oil sands industry. The book describes, in vivid detail, how humanity's increasing obsession with burning fossil fuels for practically every reason conceivable has led to our current climate crisis, which in turn is now driving the explosion of wildland megafires of terrifying size and severity.

Williams decided to interview Vaillant, and called him up. What Vaillant told him left Williams shaken. The message, based on Vaillant's extensive reporting into the Fort McMurray disaster, sounded much more dire than what city officials were telling Yellowknife residents. "What I would be doing right now is planning my escape route," Vaillant said. In Fort McMurray, officials had delayed even considering an evacuation until the last minute, then it became "how do we get everybody out of here without them burning alive on the road?"

The conversation left a profound impact on Williams. "About five minutes into that interview, I just started thinking to myself, 'Why am I still on this side of this wildfire?'" Williams said. To him, it sounded as though the Yellowknife situation wasn't that far removed from Fort McMurray seven years earlier. In fact, the fire conditions in Yellowknife were not as extreme. The city was facing moderate temperatures and relative humidity of around 30%. Fort McMurray had seen extreme crossover conditions of 32.6 degrees Celsius and humidity drop to a bone-dry 10% the day the fire overran the town. Still, it was enough to give Williams a strong sense of dread.

But some parallels were too obvious to ignore, not the least of which was how emergency officials were communicating with the public. Yellowknife officials—like those in Fort McMurray—were steadfastly refusing to acknowledge, at least publicly, that the whole city might have to flee. As of August 15, the plan was to shelter in place within the city, pulling residents back from outlying neighbourhoods to areas closer to the lake if and when the fire rolled in, housing them in sports complexes and arenas. No one appeared to be considering what could happen if an ember storm threw firebrands beyond the forest's edge and into the city proper, if—like Vaillant explained of Fort McMurray—whole neighbourhoods wound up surrounded and then consumed by fire.

Williams was convinced. It was time to leave. After his unsettling conversation with Vaillant, Williams raced to prepare the interview as both a podcast and a Q&A, and posted it to Cabin Radio's website. Then he called his partner, Liny Lamberink, and said, "Look, I think we need to go." They debated the idea for about twenty minutes, trying to decide—like many Yellowknifers—whether to trust the city's messaging and stay put, or pack up and go despite there being no formal order to evacuate. Ultimately Williams's arguments won out. They'd gather their things, drive out through the fire that night, and hopefully continue their work from outside the fire zone. They would take the risk out of the equation, their pets would be safe, and neither would have to worry about interrupting their coverage for an unplanned race to safety. They started preparing to leave. Lamberink had already prepared an emergency kit for them, a large Rubbermaid box with supplies, a first aid kit, emergency blanket, food for for themselves and their pets, and a giant tank of water.

As Williams and Lamberink loaded bags of dog food and other emergency supplies into their truck, Brookes was deciding she couldn't take worrying anymore. They needed to get out. For days she had watched as public officials offered little but platitudes and reassurances, nothing that resembled a plan as far as Brookes could see. As the fires spread closer and smoke washed into the city, it seemed to Brookes that officials were denying a dangerous reality that was increasingly inevitable.

Sterling's birth hadn't been easy, and he was delivered by Caesarian section, an intense abdominal surgery that typically requires six weeks to fully recover from. But Brookes and Purcell didn't have that kind of time. They had decided to leave despite there being no formal evacuation order in place because they were worried about newborn Sterling and the impact the horrible air quality might have on his development. They were also worried about getting caught in a last-minute traffic jam if the city did call for an evacuation. So Brookes held Sterling close as Purcell loaded up their SUV—along with their two dogs and two cats—and set out on the highway that night, knowing they'd have to drive through the fire, but with no idea where they were headed beyond that.

As they drove, one of the fears that had motivated their voluntary evacuation crept back into Brookes's mind. It was pitch-black on the highway. The smoke was worse than in town, and behind the wheel Purcell struggled to see where they were going. When they reached the fire itself, they could see flames through the trees beside the highway.

In her mind, Brookes couldn't stop thinking about scenes from Hawaii, of the stories of people getting trapped on a highway by fire and burning alive in their cars. Her chest tightened. Her pulse quickened. It was getting harder and harder to breathe. She realized she was having a panic attack, and begged Purcell to turn the truck around and drive back to Yellowknife. They got home exhausted sometime after midnight and booked a flight out for Sunday, August 20. As for their dogs, "we'll just figure it out," Claire said, with no idea yet what that would mean.

As they packed, Williams also started calling the rest of Cabin Radio's staff to tell them what he and Lamberink had decided. Most of the staff agreed it was best to leave and decided to drive out together, in a convoy for extra safety. As they prepared to evacuate, Williams also unboxed a Starlink mobile satellite receiver that had been sitting under his office desk for almost a year, and set it up wedged between their bags in the back of the truck, anchored in place by a giant bag of dog food. With the territory's only internet link now cut by the fire, Williams's Starlink would allow them to stay connected and keep reporting, even while they were on the road. Once everything was packed, he and Lamberink tried to get an hour's sleep. By about 4:45 a.m., the convoy was ready to go. They rolled out of the city and down the highway towards Fort Providence through the densest smoke Williams had ever seen.

As they headed out of the city, a small worry started growing in Williams's mind: With public trust in the city and territorial government so low, Cabin Radio had become the unofficial emergency communicator for twenty thousand people, with Williams as its very visible figurehead. Tensions had been building in the city for days, as more people began to doubt the government's ability to keep them safe. Across Yellowknife, families had been agonizing, like Brookes and Share and Williams himself, over whether to leave despite there being no evacuation order in place. Throughout the crisis, it was Cabin Radio's frank and honest tone that had earned them listeners' respect and trust. Would a public acknowledgement of Cabin Radio's self-evacuation cause its listeners to panic? What would happen if they announced they were fleeing and then ran into difficulty or lost signal en route? Would that lead people to think the road out wasn't safe?

"Am I going to freak a bunch of people out? Do I have a responsibility to explain that to people, and say 'Look, I don't like the look of this situation, so here I am.' Is that going to help other people? Is that irresponsible?" Williams decided he would shut down the station's live feed until the team reached safety on the far side of the fire, assuming—of course—that they could.

The drive out of town was horrendous. Even inside the cab of the truck, wearing an N95 mask, Williams still struggled to breathe. As they drove along the highway, the smoke closed in further until they could hardly see. Then, like a shepherd, an emergency vehicle appeared seemingly out of nowhere and pulled in ahead of Williams and Lamberink's truck. It guided them for about twenty minutes through the dense smoke and smouldering fire until they cleared the other side, when Williams was hit by an odd emotional sensation. "Okay, we're through the worst of it," he thought, "but now I have left everything behind." His house, their belongings, the radio station and all of its equipment; there was now a huge and growing wildfire between Williams and the structure of most of his life.

As they drove free of danger, Williams couldn't help but imagine how silly they'd look if there ended up being no evacuation order at all. "What if there's some almighty rainstorm, and by Friday everything's fine, the air has cleared, there's volleyball being played in the parks of Yellowknife, and here I am perched in Fort Simpson looking like a fool," he thought. He would not have to wait long to find out.

CHAPTER 6

EVACUATION

When Brookes and Purcell woke up on August 16 back in the city, the heavy smoke that had hung over Yellowknife for days had finally started to clear, but now the skies snarled with the sound of turbine engines, driving water bombers and helicopters to the fire and bringing them back again. The planes weren't new—they'd been filling up on Great Slave Lake and flying missions to the fire for days, yet now their runs appeared to be getting more frequent. The short time it was taking them to refill, drop a load of water on the fire and return to fill up again made Brookes nervous. The fire was getting closer to the town. Even with the improved conditions, it was still so smoky that Brookes could feel it in her throat. They had air purifiers running in nearly every room in the house. Though it was sweltering, they kept all the doors and windows closed.

Brookes spent the day calling friends, swapping what information they were hearing, most of it rumours. Yes, an evacuation was coming, one person insisted. No, it wasn't, another claimed.

Pets would be allowed on flights, no they wouldn't. Several of Brookes's friends also had young kids, and talking through their thinking around whether to stay or go was helpful and, in a way, calming. While Brookes's parents also lived in Yellowknife, they were travelling out of the territory at the time, but her dad had heard through the grapevine there was going to be some sort of an announcement that evening.

By 8:30 that same morning, Williams and Lamberink had arrived in Fort Providence, a tiny hamlet and major refuelling point located roughly 315 kilometres southwest of Yellowknife. It had taken Williams and Lamberink more than four hours for a drive that usually took three. Now, as they pulled into the local gas station, they realized how many other folks had made the same decision to get out early. There was already a lineup for gas in a town of barely six hundred people that rarely had to line up for anything.

While they refuelled, Williams fired up the live blog again. He let tens of thousands of people know they'd chosen to evacuate and had made it through the fire. "By this point, we were breaking web traffic records daily," Williams said. "Getting back online, I was able to tell people, 'Look, here's the decision we made, here's what the drive was like.'"

Williams and Lamberink kept on driving, reaching Fort Simpson, another three hundred kilometres away from Yellowknife, around mid-afternoon. By that point, Williams was getting messages from senior government sources that said a formal evacuation order was coming. The news caught Williams and his team in a high-stakes moral dilemma. Given the fire's proximity to the city, and how much anxiety people had about whether to stay or go, this was crucial—potentially lifesaving—news. Cabin Radio had a duty to confirm and report it as quickly as possible. Unofficially, word was already starting to leak out. Some government officials had already been telling friends and family, giving them a head start. According to internal government reports released years after the fire, some staff members had even used their insider knowledge to start preparing for the evacuation themselves before the rest of the city knew it was official.

Williams's team had reliable sources. He trusted them and considered the tips accurate, but reporting that an order was coming before an official announcement was made risked causing a panic. Williams weighed the options carefully and decided to hold off. He'd done his part by explaining his team's decision to leave, and there was nothing stopping other people from making the same one. Cabin Radio decided to hold off until an official order was made, then reported the virtual press conference in real time on the live blog just after 7:30 p.m. on August 16. Agata Gutkowska, the territory's press secretary, opened the conference sounding harried.

"I'm going to get started because we don't have much time," she said, before introducing Shane Thompson, the territory's minister of environment and climate change, alongside Yellowknife Mayor Rebecca Alty and a host of other emergency and public officials. Waiting on the line were dozens of reporters on the call from across the territory, the country and as far away as Germany. After fumbling to unmute his video microphone, Thompson spoke first.

"Unfortunately, our wildfire situation has taken another turn for the worst, with the fire burning west of Yellowknife now representing a real threat to the city," Thompson said, reading from a script. "Without rain, it is possible it will reach the city outskirts by the weekend. Facing this possibility, I am directing residents of Yellowknife, Dettah, N'Dilo, and Ingraham Trail to begin evacuating in a phased approach determined by the level of risk," he said, stumbling slightly as he rushed through the words.

Thompson gave residents until Friday at noon—less than forty-eight hours—to get out of the city. "You put yourself and others at risk if you choose to stay later," he said. "Please ensure you have enough fuel before leaving," he added. The fire was now just sixteen kilometres away.

When it was the reporters' turn to ask questions, they came fast and furious. How will the evacuation flights work? How many will there be? With only two gas stations along a seven-hundred-kilometre stretch of highway, how would people be expected to fill up their cars? What about people who don't have vehicles? What was being done to help unhoused folks in the city?

When reporter Luke Carroll from the local CBC station got on the line, he cut right to the point. "Yeah, I got *so* many questions about this evacuation strategy, but I guess I'll just start," he said, the frustration behind his words audible, even over the video call's digital crackle. "This is perhaps one of the most pivotal moments in the city's history. We're talking about the city potentially burning down. Why hasn't the public been provided any information all day today?"

Officials gave the kind of routine canned answers officials often give at press conferences. It was a logistical challenge organizing so vast an evacuation. The fire behaviour had been unprecedented. When wildfire information officer Mike Westwick got on the line, he told reporters that firefighters had been battling the blaze for over a month and losing. It had burned over their control lines time and time again.

But despite the many questions asked and despite the many officials who spoke in the hour-long press conference, there were few concrete answers.

As he wrote up the copy, Williams thought back to the last press conference he'd covered, two days and one harrowing drive through a fire earlier. In it, city officials had said the only way Yellowknife would be forced to empty was if doomsday were on its doorstep. Now they were telling residents to get out and giving them less than forty-eight hours to do it. To most Yellowknifers, the obvious assumption was that doomsday must have arrived. "We went from 'We're not having an evacuation' to 'Everybody get out of here now' in less than forty-eight hours," Williams said. "Everyone is going to lose their shit."

BROOKES WAS SITTING IN HER LIVING ROOM, GLUED TO THE press conference on TV, listening as government ministers ducked questions and deferred to their staff. She heard Thompson deliver the news as soon as it broke and felt both a sense of relief and a growing new worry. "We all felt very in the dark about what the plan was going to be," she said. "There's only one highway and one airport."

The sudden reversal from officials deepened Brookes's fears. "Do we really have a couple days? What if the wind shifts?" she worried. Everyone in Yellowknife knew that just days earlier a sudden wind shift had wiped most of the town of Enterprise completely off the map. The fears of getting trapped in a line of cars while fire swept in around them crept back into Brookes's mind. She and Purcell made a plan with his brother to drive out together, and they began frantically repacking their belongings. Brookes grabbed diapers, baby clothes, and a few sets of clothing for herself and Purcell. She forgot to pack shampoo and baby wash. When she and Purcell had decided earlier to leave as a precaution, the fact that her house might go up in flames hadn't truly set in. It had been more of an abstract fear than a real one. But now she raced around trying to pack whatever she couldn't bear to lose.

She added their passports, marriage license and birth certificates, food for the dogs and cats, and whatever cash she could find in the house. Into their compact SUV, they crammed three humans, two large dogs, two cats, a rabbit and whatever luggage they could fit. Purcell's brother had a pickup truck, which allowed them to bring a little more stuff, but even so, the family and pets together made the Jeep feel like a clown car.

By about 11 p.m., Brookes and Purcell were ready and headed back out onto the highway. This time, instead of a lonely road covered in smoke, they met an enormous line of cars already snaking into the fading late-summer evening light. At least this time they weren't alone, Brookes thought to herself.

But as they joined the massive lineups, that proved more curse than blessing. Having twenty-thousand people all trying to drive out of a city down an uneven two-lane road became next to impossible, and simple human needs became one of the biggest challenges. It took Brookes and Purcell even longer to reach Fort Providence than it had taken Williams and Lamberink. When they arrived, not only was the lineup for fuel now seemingly endless, but there were also no functioning bathrooms. The station's plumbing wasn't designed to handle thousands of people per hour, and they'd run out of water.

Some people suggested she just go squat in the woods, Brookes said, "But I physically couldn't. I'd just had major abdominal surgery." One person pointed her towards Porta Potties on the edge of the gas station parking lot, but they were already close to overflowing by the time Brookes, Purcell and Sterling arrived. Eventually she found one that was barely usable and, with no water to wash with, did her best to not touch anything.

The government had commandeered town halls and hockey rinks along the escape route to act as evacuation centres, but they were small and crowded. Brookes worried about bringing a now six-day-old baby into a "gymnasium full of germs," so they just kept driving. It would normally take around thirteen hours of driving to reach Whitecourt, Alberta—a resource town 1,300 kilometres from Yellowknife and 180 kilometres northwest of Edmonton. As they drove, however, the ETA provided by her phone's navigation app kept getting longer. Brookes kept scrolling Expedia, trying to find a hotel with a room they could stay in, growing increasingly desperate as the hours and kilometres rolled by. Having slept little the night before, and then spent an anxious afternoon waiting for news, "we were already sleep deprived before hitting the highway," Brookes said.

The situation was similar for people all across Yellowknife. People who tried to flee with such little notice wound up running out of gas along the highway, or they packed in such a panic that they left important belongings behind. Others—like Brookes and her family—crammed their precious belongings and pets into cars and set out on an odyssey they had been given hardly any time to prepare for.

Brookes and her family wound up driving for twenty-five hours straight, stopping only briefly for gas or awkward bathroom breaks. Though she wasn't supposed to be driving given her recent surgery, Brookes ended up behind the wheel so Purcell could sleep—it was safer than the alternative, she figured. As they drove through town after town, every hotel and motel they found was full.

Eventually, they reached Whitecourt at close to midnight on Friday, August 18, and Brookes was relieved to find a hotel that still had one room available. By the time they arrived, Brookes's feet were so swollen from sitting for that long, she couldn't see her ankles anymore. She hobbled up to the front desk and asked to check in. But the desk clerk at first refused to accept them because the hotel had a no-pets policy. Brookes immediately burst into tears.

"We've been driving for twenty-five hours straight," she told the clerk. "It's not safe for us to drive anymore."

"Yes, but what if the next person after you has allergies," the clerk replied, with apparent sincerity.

Shaking with exhaustion and barely six days post-surgery, Brookes managed to shoot the woman a look fierce enough to melt asphalt. The clerk relented.

When they walked into their hotel room a few minutes later, Brookes nearly started laughing. The room reeked of stale cigarettes and weed, mingled with the pervasive scent of campfire that now hung across the whole province. But sure, maybe the next guest might be upset by a little pet dander. With Sterling in a bassinet beside them, Brookes and Purcell collapsed onto the hotel bed and—finally—slept.

SHARE WAS AT A FRIEND'S HOUSE ON WEDNESDAY EVENING when the evacuation order came down. Everyone else in the household had already evacuated. As they sat and watched the press conference together, what had until then been only a theoretical conundrum suddenly turned into a very real one.

Share and her colleagues had already put together a plan that included a chartered flight to Fort McMurray for their clients. Before any of them could take it, though, she had to find them—and she knew that tracking some of them down would be a challenge. Many were unhoused and sleeping rough outdoors. They would usually access the shelter or the sobering centre throughout the day, but not reliably enough to know when exactly they'd be there. Many likely wouldn't even know about the evacuation yet. Share went to bed that night with a vague sense the next day would bring chaos, but she had little expectation beyond that. The whole city was now in uncharted territory.

The next morning, Share went to the Women's Centre and started organizing clients and staff for a shuttle to the airport. Meanwhile her colleagues mounted a physical search around the city for unhoused people. That meant driving around in the early morning hours, checking the known spots where people would typically sleep, then waking them up and trying to convince them to stand in a lineup for hours, all while helping many of them avoid alcohol withdrawal so they didn't disappear in search of a drink. At the Women's Centre, Share found herself in the midst of tricky conversations, explaining why some people were being offered a flight while others might have to wait while they were registered at a local high school or they sorted out their IDs.

By now the city's evacuation was in full swing. Public officials had urged those who could stay with friends or relatives outside the territory to do so, but there were many who would need hotel accommodations and hundreds who would need the free evacuation flights.

As thousands of people joined the growing convoy of vehicles fleeing south, hundreds more tried to register for evacuation flights, queueing in long lines for hours outside a local high school.

"The air evacuation queues were heartbreaking," one Yellowknife MLA later told authors of a government investigation into how the evacuation was handled. There was no support for the elderly or people with mobility issues, many of whom were forced to stand for hours in lineups with little food or water, and no access to bathrooms. Indigenous community members in large families were sometimes split up, and people who spoke languages other than English were left confused. "The queue management was honestly disgraceful. People were doing the best they could, but this was totally unacceptable as an operation in emergency management," the MLA wrote.

At the airport, evacuees were herded onto planes without any security screening and at times little coordination. Some planes left with empty seats, contributing further to delays. One WestJet pilot, seeing the confusion, reportedly refused to take off until all the seats on his plane were full. Some evacuees wound up boarding planes with no idea where they were heading. A woman working at the check-in counter said many of her colleagues simply hadn't shown up for work that day. They'd presumably joined the mass evacuation along with everyone else, and the airport was doing the best it could, a pattern that would play out among multiple critical organizations over the next several days.

In the rush to execute the evacuation, neither the city nor the territorial government had created direct instructions to determine who would be considered an essential worker and required to stay, and who should join the thousands of people fleeing. Deciding who should stay or go during an evacuation is a standard part of most jurisdictions' emergency planning. Workers who are responsible for the core functions of a town are often asked to remain until it becomes unsafe for them to do so. This can include people who operate a town's utilities and water systems, for example, or those who staff its hospital or medical facilities. Someone must be around to make sure firefighters have water, for example, and to treat them if they're injured. Essential workers often include the mayor and key bureaucrats. And the designation almost always includes police officers and anyone working directly to manage the emergency itself.

The confusion left government workers in Yellowknife grappling with a moral dilemma. If they stayed but weren't supposed to, they could tie up critical resources or hamper firefighting efforts. But if they fled and weren't supposed to, there might not be enough people to keep the lights on and the emergency management system running.

"What am I? An evacuee? An essential worker? An evacuee who continues working but is not deemed essential?" asked one city employee who worked in emergency management. Even workers for one of the territory's helicopter companies, who had for weeks been supporting firefighting efforts directly by flying crews to the front, didn't know whether they should stay or go.

At Stanton Territorial Hospital, things were even worse. For firefighting efforts to continue, firefighters needed an operational trauma centre in the case of any accidents or injuries. But hospital medical staff were evacuated by mistake and had to be flown back to Yellowknife to keep the hospital's emergency room functioning. And while most of the hospital's patients had been flown out, the most high-risk ones—those needing round-the-clock care or patients on ventilators or life-support—were harder to evacuate.

At one point, a military flight scheduled to take high-risk patients from the hospital, along with the nurses who cared for them, was "aborted" with little explanation. "We don't have answers," one frustrated nurse told Cabin Radio. "When will the military pick up the patients? When will everybody be safely evacuated?"

At the airport, there was more confusion when Share's staff started arriving with cases of beer. Once Share explained that the booze was supplies for the managed alcohol program, airport staff allowed it. But it's one thing to bring cases of beer with you. It's another to be handing out drinks to evacuees waiting hours in a lineup. Clients on the program would regularly get a drink every two hours, but it took much longer than that to arrange and board the flight. Share and other YWS workers wound up handing out drinks to some of their clients in the waiting lounge. "We were basically doing the managed alcohol program on the fly," Share said.

Some of Share's clients who'd agreed to fly out didn't show up, and others who hadn't registered did. At one point Share was rewriting the passenger manifests by hand with a pen. Some people lacked any sort of government-issued ID, but despite fears this could leave them stranded, Share said the airline staff accommodated everyone. When they all finally boarded the plane and lifted off, there were some empty seats and Share worried they'd left more people behind than they should have. She tried to reassure herself that at least they'd managed to get dozens of the most vulnerable people out amid the chaos.

When their charter flight arrived in Fort McMurray, they were met by a series of school buses that ferried people to an oil sands work camp. Many of Share's clients would end up staying for the next three weeks. By the time they landed and despite their best efforts, many clients had missed one of their scheduled drink windows and were at risk of going into withdrawal. For people with complex needs, in the midst of an already stressful emergency in a strange city, this had the potential to create its own mini-crisis if clients started wandering away in search of a drink. A registered nurse who works with Share's organization focused on getting the scheduled drinks flowing again, while Share set about finding somewhere more formal to host the alcohol program.

The work camp was a dry facility, and unloading cases of Molson raised a lot of eyebrows. Just as she finished solving that problem, Share's phone started buzzing with calls and texts from other clients still back in Yellowknife, frustrated at having waited all day in lineups only to be told to come back on Friday, or to go to a different evacuation centre. Some folks wound up on flights to Calgary or Edmonton, so Share started working the phones to connect them with critical local support. For some of Share's clients, Yellowknife itself was the "big city," already a significant adjustment from their much smaller home communities. Landing in a place like Calgary with little support and no family or community connections created a significant risk. Alcohol withdrawal can cause seizures and can sometimes be fatal. Bad actors of all sorts, from drug dealers to human traffickers, are known to prey upon vulnerable populations, particularly at times of crisis and dislocation from community. Share is too humble to admit it, but given the maelstrom of risks and challenges faced by her clients, the charter flight with its managed alcohol program likely saved lives.

THROUGHOUT THE DAY, CABIN RADIO KEPT ITS LIVE FEED RUNning, tracking all the updates in a single place. It seemed that everyone, from those still inside Yellowknife to the thousands of people strung somewhere along the 1,300-kilometre evacuation route, was relying on their reporting. Between 6:30 a.m. and signing off at midnight, the news team wrote nearly 199 updatesa new post every five minutes.

Williams and his team weren't just covering the situation in the city; they were trying to stay abreast of a wildfire crisis that was still unfolding across the rest of the territory as well. An entry from their live feed at 11:20 a.m. speaks volumes: "**EVACUATION ORDER FOR KAKISA** dear god," Williams wrote. "All residents of Kakisa evacuate immediately, oncoming wildfire. Evac centre is here with us in Fort Simpson. Get out of there now."

While they covered these and other distressing developments, a key focus of Cabin Radio's live feed was highlighting important information for evacuees fleeing along the highways, including which evacuation centres were open, which ones were full, where to find gas, and more. When the Alberta town of High Level opened its sports complex, baseball diamonds, parking lots and even some private residential properties for evacuees to camp on, Cabin Radio broadcast it.

Readers sent in photos showing where the worst smoke was along the route; what was happening at the evacuation centres they reached; and even shots of themselves in vehicles loaded high with kids, pets and belongings. Williams responded in kind, often with his trademark humour. At 2:15 p.m., in an update telling evacuees the gas station in Fort Simpson would remain open through the night, he also added, "I went into the Northern [Store] just now and a Pizza Hut pizza was sat there in the warming stand at the front counter, I swear to God bathed in the light of a million suns. I cradled it in my arms like my first-born and have now, at last, eaten something today."

At 2:45 p.m., the territorial government announced that evacuation flights for the day were full, and the hundreds of people still in line would have to come back the next day. "We know this is deeply frustrating for those who have been in line for several hours and who will need to line up again tomorrow," the official statement said. But it turned out to be wrong. Flights continued almost every hour until late into the night, past the next day's evacuation deadline and well into Saturday.

Amid the confusion, many Yellowknifers started organizing themselves, arranging carpools and rides for folks who wouldn't—or couldn't—wait in the airlift lineups for hours on end. One particular challenge was the sheer number of pets in town. Virtually every household had at least a dog or cat, many had more than one. One woman evacuated with a pet boa constrictor; someone else fled with a pet turtle in a cardboard box. Another drove out with her car loaded full with cages of her friend's ten pet budgies. With the vehicle full of birds and no room for the friend, the birds' owner took an evacuation flight.

Through all of this, Williams also became a sort of unofficial spokesman for the territory at large, doing interview after interview for what seemed like every TV news network on the planet. He did the first one at 6 a.m. from the front porch of his hotel in Fort Simpson, in a rumpled hoodie and running on next to no sleep. After his second on-camera interview, Williams realized—somewhat ruefully—that he had not brought enough clothing, and certainly nothing that befitted (in his words) a "supposedly professional journalist" being interviewed live during the biggest story of his career. In TV interviews that aired nationally and went viral online across the globe, Williams had to alternate between a blue Team Northwest Territories hoodie and a grey Roots one, the two nicest pieces of clothing he had with him. "I packed like I was going camping for a long weekend," he said.

In between the interviews, Williams and the rest of the Cabin Radio crew kept reporting on the fire, the evacuation, developments of all sorts. But what resonated the most with their audience during those first few days was their live feed. "We kept asking our audience, 'Tell us what your experience was so we can help the next people out,'" Williams explained. "How was the smoke along the road, how much gas was there at the gas station, how long was the line? Could you see fire? Have you got to Alberta yet?"

This vital crowd-sourced information provided exactly what stressed-out evacuees needed, and what government itself was unable to provide. Along with giving fleeing residents crucial information, Cabin Radio's coverage also functioned as a sort of community pressure release valve. People called or wrote in with all manner of information, some of it humorous, much of it critically important. Williams's team kept this up for weeks after the evacuation, constantly rebuilding their webpage to foreground the most important, useful information. Thousands of Yellowknifers came to rely on them for answers about everything from where the fire was on a given day, to how to access evacuation support, where to find shelter, and when they might finally be able to go home.

Through it all, Williams and his team held to their core mission of an outlet with a warm, friendly and adventurous spirit that mirrored Yellowknifers themselves. Their coverage didn't sensationalize or catastrophize the situation, but it did fill a critical information hole left by the collapsing government information systems.

AS RESIDENTS SCRAMBLED TO FLEE, THE FIRE CONTINUED BURNing roughly 16 kilometres outside of town. Yellowknife had failed to do any fuel reduction or FireSmart work for the past two years, and now the race was on to get last-ditch defensive lines constructed before the fire reached the city. Teams of local contractors from roughly twenty companies, along with seventy-five volunteers who stayed behind, went to work on a series of enormous fire breaks around Yellowknife. Some of the work had started on August 7, but when the fires breached containment lines west of the city on August 15, construction took on a fresh and desperate urgency. Workers put in ninety-hour weeks coordinated through ad hoc WhatsApp group chats. They used heavy equipment like bulldozers and excavators to plow away hundred-metre-wide swaths of forest around key areas of the city. Industrial contractors from the city's many mining companies cobbled together more than twenty kilometres of pipeline and hoses to feed huge cannons that blasted out water to soak the fire break's leading edge. The work was so last minute, there was no time to order the necessary pipes and fittings. They fused together what they could scrounge from their companies' existing inventories and from around the city. As one worker told Cabin Radio: "Everybody's truckin' up here. There's nobody standing around, and we're going to be ready for this thing."

The majority of the work was accomplished in an incredible five days. As one local contractor told Cabin Radio, the project was—if somebody had actually sat down and planned it—probably months of work. He described what happened as competitors putting down their swords to build a massive communal shield. "It's a level of cooperation among local contractors that I've never seen," he said, "and I've been here over thirty years."

As impressive as the fire breaks were, they never guaranteed protection. Modern wildfires can easily overcome even the most enormous water cannons blasting thousands of litres per hour. A common misconception about fire breaks is that they are capable of stopping a wildfire in its tracks. It's similar to the frequent belief that water bombers put out fires. Neither is true—what both tactics can do is help to cool down a fire's behaviour, slow its spread, and reduce the amount of embers it launches into the sky.

But even that only buys protection across the reach of the water cannon and immediate width of the fire break itself. Modern wildfires can launch hundreds of thousands of embers the size of baseballs five kilometres or more ahead of the fire front. In the same hours that Yellowknife workers raced to complete the city's defences, the Adams Lake wildfire in BC's North Shuswap demonstrated just how much damage a monster wildfire can wreak, easily blowing over the region's last line of defence and raining embers across dozens of square kilometres into Meadow Creek, Scotch Creek and Celista. If that happened to Yellowknife, even the city's robust new fire break would be powerless to stop it.

Luckily, the city's last-minute defences were never tested. On August 21, five days after officials ordered Yellowknife's full evacuation, heavy rain fell across the fire zone and dampened the flames. The rain gave firefighters the opening they needed to attack the fire directly for the first time since its ignition nearly a month before.

After their night of much-needed rest in Whitecourt, Brookes, Purcell and Sterling hit the road again and met up with family in Peace River. They left the rabbit and the cats with their family, making a little more room in the cramped Jeep. From there, they drove to Edmonton and caught a flight to BC to stay with Brookes's family. It was jarring to be suddenly in a big city, so far from the confusion and fear they'd just fled, surrounded by strangers just going about their days. At the Edmonton airport, Brookes remembers getting strange looks from other passengers. "I just felt like so many people were under the impression that we were going on vacaltion with a twelve-day-old baby and two eighty-pound dogs, like we were going to have a good time and collect seashells or whatever." She proudly wore her favourite hoodie with "Yellowknife" emblazoned across the chest in block letters so other travellers would know where they were from.

After three weeks away from home, Yellowknifers were finally allowed to return at 11 a.m. on September 6. For Brookes and Purcell, and many others who'd fled the city with no time to prepare, the return to Yellowknife was as much of a logistical nightmare as leaving had been a frightening one. Through no small amount of juggling cars and dogs and cats around BC and Alberta, the family was finally able to fly from Vancouver back to Edmonton and from there a flight back to Yellowknife.

For Share, getting all of her organization's clients home again was at least a little more straightforward. After getting everyone settled at the work camp near Fort McMurray, Share and her partner were eventually able to fly to Edmonton and take a few days off, which she spent mostly catching up on sleep in a hotel room. From there, they headed to Banff, which had started offering free camping for evacuees, and Share worked remotely fielding calls from staff and clients, trying to find out how long everyone should expect to be away from home. There were, unsurprisingly, few answers to be found. When word came that it was finally time to go home, Share and the rest of the YWS staff essentially ran their evacuation process in reverse, this time having had the benefit of practice.

Williams and Lamberink were some of the first residents back in the city. Williams parked his truck in front of the Cabin Radio office, balanced his laptop on the hood, and recorded a video under a clear blue sky. His home, and the city itself, had avoided catastrophe. "It felt genuinely great," he said.

In some ways, for Williams and the Cabin Radio crew, the story of the fire never really ended. The team continued its dogged coverage during their weeks away from Yellowknife and after they got home, turning from breaking news updates to analysis and investigative work holding territorial and city officials to account for how the evacuation was mishandled. They kept it up for years, giving the issue its own dedicated landing page on the Cabin Radio website. Reporter Emily Blake's work was particularly impressive, dredging up a trove of internal government emails that gave readers a glimpse at the dysfunction behind the governments' curtains. Years after the fire, they were still breaking stories about the crisis and its aftermath.

WHILE YELLOWKNIFE ESCAPED THE FLAMES, IT DID NOT ESCAPE without scars. For many residents, the chaos of the evacuation was itself a form of trauma. There's an argument that the city and the territory's handling of the crisis deepened that trauma, in part because it did not give people any meaningful agency. By repeating a mantra of "everything is fine" right up until it wasn't, public officials denied residents the ability to decide for themselves whether to stay or leave.

In the years following the NWT wildfires, the City of Yellowknife and the territorial government commissioned three significant reports investigating how the municipal, territorial and Indigenous governments handled the crisis. Each would largely confirm what most Yellowknifers already knew in their guts; it had been a mess, and a lot of things had gone wrong that didn't have to.

The first, released in June 2024 by global consulting firm KPMG, examined the City of Yellowknife's response to the fire and found major flaws. City staff had done little training or preparation for an emergency of this scale, the report found. The city had done no work to reduce wildfire risks around and inside the community for two years leading up to 2023. It also found that the sudden effort to build massive fire breaks and sprinkler lines around the city diverted critical resources away from planning for the evacuation.

More than 80% of Yellowknife residents surveyed said they were only able to access "low" or "very low" levels of relevant information about the wildfires, the city's response, and the evacuation procedures in particular. Many residents told KPMG they would have preferred to be told point-blank there was no plan at all, rather than be told there was one only to later find out that was essentially fiction.

A similar review, commissioned by the territorial government from Calgary-based consulting firm MNP examined how the territory's Ministry of Environment and Climate Change, which oversees the wildfire service, responded to the crisis and was equally blistering. It was released in July 2024 and looked at the response not just to the Yellowknife fire, but to fires across the territory that year. It found glaring holes in the territorial government's preparedness, including inadequate fire behaviour modelling, outdated data tools and a lack of appropriate training for fire behaviour specialists, the very people whose job it is to predict how fires will move on a landscape.

It also found bad coordination between all three levels of government. There was confusion about who held which roles during the crisis, how incident command systems were meant to operate and who held responsibility for fighting fires outside community boundaries. The MNP review even found that the territory often lacked enough trained staff for "complex" operations like indirect attack, a basic firefighting strategy employed by wildland firefighters the world over.

A final accounting was released in May 2025 by consulting firm Transitional Solutions Inc. It found that—amid all of the chaos of unclear roles, responsibilities, competing jurisdictions and deviations from standard operating procedure—the territory did, in fact, have an emergency management plan all along. It had been created in 2018, and it "should have worked." The only problem? Many people didn't know it existed and those who did largely didn't know how to implement it—the territorial government had not run any significant training or tabletop exercises to practice.

The other major theme that echoes across all three reviews:

Both the city and the territory now face a years-long effort of regaining public trust that was shattered by their joint mishandling of the evacuation process. In the end, more than nineteen thousand people fled the city either down a bumpy two-lane road, or in rushed and disorganized airlifts. It's lucky that nobody died in the process.

Looking back, Share believes that the fault doesn't just lie with government. There had been fires burning all summer across the country and no shortage of stories about chaotic, dangerous or rushed evacuations. "We should have thought about this too. A lot of it came down to just thinking, well, it's Yellowknife. It's not going to be evacuated," she said. For many people in the social work community, Yellowknife is the city people come to, not the city people flee. Other outlying communities across the Northwest Territories are much more familiar with evacuations, and not just from wildfires. Spring floods are also a common trigger. And in those cases, Yellowknife has always functioned as the safe haven. Shifting that kind of thinking can take a monumental effort.

One of the strongest lessons from the crisis that sticks with Share is the importance of giving people choice and agency whenever possible, especially during an emergency that has, by definition, taken many of those choices away. There's a school of thought that conceives of trauma at its most elemental as a loss of control. Throughout the evacuation, Share and her team took care to never force anyone to do anything. They created as many workable options as they could and let their clients choose what they wanted. Not everyone accepted the help, but Share is convinced those who did benefited from being able to choose it freely. It's a philosophy that stands in contrast to most governments' approaches to emergency management, which tends to dictate to people through orders and commands—not unlike how officials treated local residents in BC's North Shuswap.

One of Williams's chief frustrations echoes what all three major reports examining the crisis had found: city and territorial governments so unprepared that they couldn't even use the three-tier emergency alert systems they did have in place. The first evacuation alert Yellowknife received was only for the outlying neighbourhoods on the city's west—the ones the official plan contemplated possibly evacuating, but only to sites within the city itself. That alert was upgraded to the whole city mere hours before the evacuation order itself was announced.

Williams said he still can't understand how it seemed so many people got caught off guard. "It's not like nobody knew . . . that this fire was there," Williams said, but officials seemed unwilling or perhaps unable to imagine that the worst really might happen. It's in that refusal to admit reality, likely even to themselves, that Williams and many others see the strongest parallel between Yellowknife and Fort McMurray, Slave Lake, the North Shuswap and any number of other foreseeable wildfire disasters Canada has been through.

In October 2025, the territorial government released its response to the TSI report's recommendations. It fully accepted about half of them, including "exploring how to better manage communications," on social media platforms and between various levels of government. The territory committed to more clearly defining who is considered "essential" during an emergency and figuring out how to keep the territory running if Yellowknife itself is forced to evacuate again.

The territorial government also largely agreed with many other recommendations but, to borrow a phrase from Williams's coverage, "In six instances, the word 'agrees' was closely followed in the GNWT's response by the word 'however.'" In some cases, the territory said the work to implement the recommendations had already happened, and in others that it was not the territory's responsibility.

Incredibly, however, the territory refused to accept what was perhaps TSI's strongest recommendation of all: a call to create a stand-alone emergency response agency that could take the lead in handling future large-scale emergencies. Instead, the territory decided it would continue to rely on what it described as an improved version of the existing local, regional and territorial emergency management teams. These teams pop up only when needed and make up the very system that TSI's report found had failed so completely during the wildfire. "Given the NWT's small population, limited fiscal and human resources, and the infrequency of large-scale emergencies requiring sustained territorial-level response, a separate agency would be costly, duplicative, and difficult to staff," the territory's official response said. "Maintaining a dedicated agency year-round—despite the episodic nature of emergencies—would not represent an efficient or scalable use of government resources."

Reading those quotes myself that October, I was shocked. I couldn't help but see echoes of the dozens of similar government failings after fires across the country. Every Canadian province has a stand-alone emergency agency. Even the Yukon, with a similar population size and facing much the same economic restraints as the NWT, has one, as does Nunavut. The NWT, almost half of which is covered in one of the most fire-prone forest types in Canada, would remain the sole holdout, the government had declared.

How are we still this short-sighted? Why do we keep making the same mistakes, failing again and again to learn the lessons Mother Nature is trying to teach us in increasingly emphatic tones. It was as if, after all the chaos of Yellowknife's evacuation and the destruction of the entire community of Enterprise, the territorial government still couldn't see just how lucky it had been.

Nobody knows for sure where the next major wildfire will strike, but one thing is certain. When it comes to wildfires, Canada keeps getting lucky. So lucky, perhaps, that each near miss just deepens our complacency, our sense that *it won't happen to us.* When disaster does strike and wipes dozens or hundreds or even thousands of homes off the map, we grieve those losses as some sort of tragic aberration, an unpredictable act of God we could not have avoided. We see only the ashes themselves, and not how much worse it could have been.

These fires are not aberrations. They are no longer "infrequent." They are our new reality and will continue to be so for decades to come. Less than a year after Yellowknife's chaotic evacuation, and the North Shuswap's shocking devastation, both would happen again, this time in the crown jewel of Canada's Rocky Mountain parks. The impacts would leave hardened wildfire veterans astounded and shaken.

CHAPTER 7

CITIES ON FIRE

As I thought about the frustration in Brookes's voice as she described her impossible drive to safety from Yellowknife, and about all the years I'd spent investigating the North Shuswap fires, I came to two important realizations. The chaos of the Adams Lake fire showed me that, faced with extreme fire behaviour and seemingly ever-longer seasons, Canada's wildland firefighting systems were becoming quickly overwhelmed, putting firefighter lives increasingly at risk. At the same time, the failures of NWT officials to adequately plan and prepare for their own extreme wildfire season mirrored the dozens of unmet recommendations from other wildfire after-action reviews I'd read. At least four previous reports investigating destructive fires going back to the 1990s have spelled out how governments can better prepare for more extreme fire seasons, but they have mostly gathered dust on bureaucrats' shelves. We have, in fact, known for decades how to address the growing threat of interface wildfires and the urban destruction they can create. We just keep allowing ourselves to not do it.

In British Columbia, investigators reviewing the 1994 Garnet Fire in Penticton, BC, wrote that fire should have been a "wake-up call" to Canadians living in what we now understand to be the wildland-urban interface. They wrote extensive recommendations about how to reduce the risk, but virtually no one listened to them. Then it happened again in 1998, after a fire in Salmon Arm, BC, destroyed forty buildings in a subdivision on a mountain above the city.

The seventy-eight thousand hectares that burned across BC that summer seem almost negligible against thousands of homes and millions of hectares burned in recent fire seasons, but at the time, the 1998 wildfire season was BC's—and to a large degree Canada's—first real experience with a bad urban wildfire. Reading these reports decades later feels alarmingly like reading one written just last season.

The Salmon Arm fire exploded after a cold front passed through, bringing hurricane-force winds that water bomber pilots faced with the same bravery as Allied air crews over Germany in the Second World War, according to a BC Ombudsman report into that fire season. "Flying large aircraft in these conditions, with hundred-kilometre-per-hour gusts and violent turbulence from this huge fire, was nothing short of heroic," the report read.

In a 2001 report the province's auditor general urged the government to improve hazard mapping of the wildfire risk across the province, encouraged communities to take steps to "mitigate interface fire risks" through landscape-level fuel treatments, improve wildfire safety and training for local volunteer firefighters and encourage communities to practice responding to bad urban wildfires.

Two years after that report was issued, the Okanagan Mountain Park fire destroyed 238 homes in Kelowna, BC, amid what was (yet again) the worst wildfire season on record at the time. Another report was commissioned, this time authored by former

Manitoba Premier Gary Filmon, with input from dozens of leading experts in the field of wildfire prevention, mitigation and suppression. It included virtually all the same recommendations as the 2001 report, but in far more detail. One of its signature calls was for more work to reduce the fuel loads in our forests and, in particular, more use of intentional prescribed fires outside of wildfire season to help protect communities. According to the Filmon report, more than eight hundred thousand hectares of British Columbia were at either moderate or high risk of a devastating urban wildfire and needed urgent fuel reduction work to reduce the threat. It should have become the seminal document guiding how Canadian communities respond to the growing wildfire threat. Instead, it sat on government shelves.

Fourteen years later, the 2017 wildfire season wreaked havoc across BC. More than four hundred structures were destroyed, and twelve thousand square kilometres of forests were burned. The provincial government decided the thing it most needed was another government report, so it commissioned Shxwetélemelelhót, a hereditary chief of the Sq'ewá:lxw First Nation (whose English name is Chief Maureen Chapman), and former Education Minister George Abbott to write one.

Their report, which examined the twin disasters of the 2017 spring floods and devastating wildfire season, was released in 2018. Bonus points if you can guess what it found.

"Fourteen years after the Filmon report was completed, hundreds of large and small communities across British Columbia remain vulnerable to wildfire," Chapman and Abbott wrote. "Though much good work has been done in response to the Filmon report recommendations, not nearly enough has been accomplished to meet the magnitude of the threat our province faces today."

The biggest difference between 1994 and 2025 is how often these catastrophes now happen. The 2017 season was—at the time—the worst wildfire season in BC history, but it's a record that stood for mere months. The 2018, 2021 and 2023 fire seasons all attained that same infamous title one after the other. The 2025 wildfire season was the second-most destructive in Canadian history.

Most critically, Chapman and Abbott found that, fourteen years after the Filmon report's urgent call for action, less than 10% of those eight hundred thousand hectares had been successfully fuel treated. The estimated cost to treat the rest, the authors said, was roughly $6.75 billion. Chapman and Abbott's recommendations echoed those of the report authors before them: more fuel treatments, more prescribed burning, better training for local firefighters, better community emergency planning.

The sum of these recommendations have remained unchanged for so long that after 2023's extremely bad season, British Columbia Premier David Eby decided that the province did not, in fact, need another report or inquiry. It would only waste money and repeat what we have known for decades.

SO, IF WE CAN'T (OR WON'T) PLAN PROPERLY FOR THESE FIRES, and we can no longer simply water-bomb them out of existence, what can we do? Finding the answer, I was learning, meant reshaping how we think about fire in the first place.

One warm July afternoon, I was sipping Grapefruit Radlers with a friend on a grassy slope above Vancouver's seawall, looking out across the water at the North Shore Mountains. We talked about the horrifying Los Angeles fires that had torn through Altadena, Malibu and the Pacific Palisades in January 2025 with an astounding speed and ferocity. In the space of hours, the Eaton and Palisades fires burned more than 150 square kilometres combined and destroyed a staggering 16,200 buildings. They were two of the worst urban fires in living memory, leaving thirty-one people dead in their wake. Public officials were thankful the death toll wasn't far higher.

Gazing across Burrard Inlet's glittering waters, my friend was quiet for a moment. Then he tuned to me. "Do you think that will burn in our lifetime?" he asked, pointing to the city of North Vancouver, with its rows of residential streets climbing from the shoreline up mountainsides thick with trees. He and his wife had recently bought a house in one of those neighbourhoods. I could hear the subtle tightness in his voice. The answer—unfortunately—is probably "Yes."

Vancouver's North Shore is some of the most sought-after WUI real estate in the world. From the lookout at Stanley Park's iconic Prospect Point, you can see some of the priciest homes on the planet. Since its founding in 1907, North Vancouver has grown into a mecca for outdoor types who want the best of both the city and the wilderness. But nestled against the mountains for which it is named, North Van is also a perfect platform for understanding the risks of a bad WUI fire. In April 2024, I followed a crew of firefighters practicing how to fight one and glimpsed for myself what the worst-case scenario might look like.

A YOUNG FIREFIGHTER GRABBED THE HOSE, UNRAVELLING IT from the back of his wildland truck, and charged down the driveway. Somewhere behind this home, set deep among the towering fir and cedars of North Vancouver's Mount Seymour, was a spot fire—one of dozens of hypothetical fires that ignited in this neighbourhood by a pretend ember storm. A crew of young BC Wildfire Service firefighters were now tasked with finding and knocking them down before the embers could spread.

The firefighter hauled the heavy hose, slung over his shoulder, down a tight alley between the house and its neighbour's fence—

little more than a metre wide—and into the backyard, followed by a crew mate.

"There!" his partner shouted, pointing to a flash of orange among the bushes. Together they maneuvered the hose further around the corner, hauled it up through the home's garden and into the trees behind. "Give me water!" one of them shouted into his radio, and the hose snapped taught. A jet of white burst out of the nozzle, knocking an orange traffic cone on its side. It landed in a small puddle with a plop.

The drill, which the crew had executed flawlessly, was part of an annual wildfire resilience conference that takes place in a different community each year in BC. In 2023, North Vancouver's Capilano University played host, and it was the perfect venue for considering a future where wildfires invade cities with increasing frequency.

As fires in the wildland urban interface become more frequent and more severe, the traditional lines between wildland and structural firefighters are starting to blur more often, just as they had in the North Shuswap when Ty Barrett's and Roy Phillips's crews found themselves facing rank 5 fire exploding out of the forests near Scotch Creek. The crew I was following for the training scenario was a mix of wildland and structural firefighters, doing the kind of cross-training that experts and government reports have long called for to address this worlds-colliding scenario. They were practicing a tactic known as bump-and-run, where a crew will work their way through a neighbourhood on foot, like infantry dismounted from their fighting vehicles. The fire truck itself crawls along at a walking pace, with hoses already deployed and strung loosely from doors or hooks at its rear. Firefighters can then run the hoses out quickly and race into backyards or alleys between houses, knocking down spot fire targets as they are called in, before moving on to the next target.

North Vancouver's most recent wildfire protection plan

looked at the number of "extreme" fire danger days in the area between 2002 and 2018. It found an average of only three "extreme" days per year, and an average of nineteen days with a "high" danger rating. That seems relatively low risk, but under bad fire weather conditions, a major fire here could easily become a nightmare: Multi-million-dollar houses are set close together in neighbourhoods housing thousands of people and nestled in among Grouse and Seymour Mountains' densely packed western hemlock, Douglas fir and red cedar forests. It's often hard to see the houses for the trees. The fact that a bad fire so far hasn't ignited in North Vancouver on one of these "extreme" days is mostly down to luck, but the odds are shifting.

A 2023 paper authored by Leona Shepherd and Mike Flannigan from Thompson Rivers University, along with the University of Alberta's Dante Castellanos-Acuna, modelled the expected increase in "potential spread days" across British Columbia's Wells Gray Provincial Park, on the edge of the Rocky Mountains west of Valemount. These are the days on which fuel buildup, crossover and fire weather align to create ideal conditions for a fire to grow significantly. Researchers found that by 2071, the number of expected potential spread days will more than double, rising to an average of sixteen across the province. They also found that both the ninety-fifth percentile Initial Spread Indexes (a numerical rating of how fast a fire is expected to spread) as well as the Fire Weather Index (a numerical rating of expected fire intensity) will increase even more. In other words: Fires will become twice as likely to start and twice as likely to spread explosively when they do—just as the Adams Lake fire did on August 18 at Scotch Creek. The Wells Grey study isn't a perfect comparison to North Vancouver, because its spruce and pine forests tend to be naturally more fire-prone than the coast's lush temperate rainforests. But if the research bears out across BC's other forest ecosystems, the potential danger will extend even to an historically wet ecosystem like the North Shore mountains. In fact, we're already seeing evidence of this trend in action.

For decades, conventional wisdom has said that wildfires in BC's coastal forests are small and infrequent. The damp, rainy conditions that spur lush rainforest growth make it difficult for fires to spread quickly despite the abundance of fuel. Initial attack crews on the coast have a relatively easy time getting to fires while they're small and while they can drown them. At least, they used to.

The Mount Underwood fire kicked off on August 11, 2025, burning on a mountainside outside Port Alberni, BC, in the centre of Vancouver Island, which was in the midst of a years-long drought. Despite long-standing expectations that coastal fires don't burn big and hot like the drier forests of the BC interior, Mount Underwood raged to more than 3,500 hectares, easily outpacing initial attack firefighters and displaying rank 5 behaviour rarely seen in BC's supposedly "wet" coastal forests—until now. Fire experts pointed to it as evidence that long-standing assumptions about coastal forests being safe from fires are crumbling.

Many of the houses in North Vancouver are cloistered by decadent cedar boughs that lean in close, wrapping homes in a sense of quiet forest solitude. They tower sometimes thirty metres over houses and spend most of the year dripping with rain or foggy dew. But on an extreme fire weather day with high temperatures and low humidity, those trees could become towering columns of fuel. For firefighters trying to push back hungry flames in among the homes, it would be like trying to fight a wildland megafire from the inside out. As I walked along the roadway photographing the firefighters' training, I couldn't help but picture what this neighbourhood would look like if it burned on a bad fire weather day.

In my mind, the towering red cedars above me turned into thirty-metre pillars of flame, showering whole city blocks with sparks. Rivers of embers, driven by eighty-kilometre-per-hour winds, washed down the streets past firefighters struggling to defend homes at the edge of the forest. Decorative junipers ignited behind them as I passed, sending out jets of super-heated gases, melting the soffits, siding and eves of a house. Those same gases preheated the trees, houses, cars and water ski boats beside them until they too erupted in flames, further driving the inferno. Fire and embers licked hungrily up into another house's rafters, and moments later the roof erupted before collapsing in on itself, taking the whole house with it in a gush of smoke and toxic fumes, pumping out more embers. Down the street an ATV caught on fire. Propane tanks started to explode. The narrow, winding streets became clogged with traffic. Panicked drivers veered over curbs and across lawns, only to find themselves boxed into one cul-de-sac after another. The Iron Workers Memorial Bridge became a traffic jam as terrified drivers crashed into each other, cutting off one of two vital routes to safety as the mountain behind them continued to burn.

When a wildfire reaches a town, the real danger isn't from flames themselves; it's from embers. Towns, like homes, tend to burn from the inside out. A wildfire primarily spreads one of two ways. The first is through radiant heat. The flames of the existing fire preheat the adjacent fuels until they reach the point of combustion. In this way, fire can move from one blade of grass to another, from bush to bush or tree to tree, all without any wind to accelerate the process. Under these conditions, a well-organized surface fire will advance at roughly a walking pace or even slower, crawling across the ground. Add wind to the equation and the flames will bend, pushing fire-heated air forward, preheating fuels in the fire's path that much faster. Generally speaking, the higher the wind speeds, the faster the fire's rate of spread.

For the fire, this works great as long as there's a consistent continuity of fuels. Fire burns through densely packed fuels faster because it takes less time to transfer the radiant heat. As the distance between fuels increases, the fire slows down because it takes longer for that heat transfer to happen. If the distance between fuels is far enough that the fire can't sufficiently preheat them at all, the fire will likely stop spreading. This is why it's possible to stop an escaping campfire by simply dragging your boot through the dirt around it to create a rudimentary fire guard. In the absence of wind, even a vigorous wildfire can be contained by a dozer guard, a road, a creek or a swamp—anything that breaks the continuity of fuels.

The second and more dangerous way a wildfire can spread is through ember cast. Anyone who's ever dropped a juniper bough on a campfire understands the basics: Burning bits of debris are carried by the convective force of the fire's updraft, lofted into the sky to settle . . . somewhere else. That "somewhere else" depends very much on the wind. And since the convective force of a fire can also create its own winds, it essentially gives a wildfire the ability to launch burning missiles far beyond the reach of its radiant heat alone. Ember cast is what causes whole towns, to burn down.

This is what happened in Los Angeles in 2025; in Lāhainā and Halifax in 2023; in Lytton, Fort McMurray, Slave Lake and twice in Kelowna, all since 2003. What began as wildfires became something far worse: *urban conflagrations*. It's what could easily have happened to Yellowknife if rain and good luck hadn't halted the blaze. Under the right (or perhaps wrong) conditions, it could happen in places once thought impervious to fire, like the temperate rainforest that is North Vancouver.

THE POINT AT WHICH WILDFIRE BECOMES AN URBAN CONFLAgration isn't just the moment homes start burning. Wildfires are perfectly capable of destroying individual houses on their own if the flames or embers from the forest can reach them. An urban conflagration, however, begins when fire starts spreading directly from one structure to another. When homes are spaced close together, the radiant heat from one is often enough to ignite its neighbour. This triggers a chain reaction. The fire burns through a block of homes just like it would through a dense stand of spruce or cedar until that continuous chain of fuel is broken, whether by a street, or a river or a fire guard. But just as a wildfire jumps a river by launching embers on the wind, an urban conflagration can also spread when burning houses collapse. When they do, they send forth a geyser of embers that wash over the whole neighbourhood, landing in gardens and on doormats and anything else flammable, igniting spot fires that spread until another house is consumed and the process repeats.

The difference between wildfire and urban conflagration isn't just rhetorical. Each requires us to think differently about our own responsibility for how we prepare. Thinking about fighting a wildfire suggests we need legions of wildland firefighters to battle them and promotes visions of an army of Nomex-clad twenty-somethings facing down the blazes somewhere "out there" in the forest. We imagine squadrons of water bombers flying to the rescue to douse flames long before they reach our homes. Wildfires, we think, are a distant problem for someone other than us to solve.

Thinking about urban conflagrations on the other hand, about our homes themselves as fuel, puts the emphasis where it needs to be: on us. It forces us to recognize that building homes in the middle of a fire-prone landscape means we have to take some responsibility for protecting them.

In the North Shuswap, hundreds of cottagers, retirees and long-time locals built their dream houses nestled deep into spruce, pine and fir trees on steep slopes crowded along the shoreline. Many of the properties have that quiet, alone-in-the-wilderness feel to them, despite being quite close to their neighbours on hillsides studded with homes. This, all in an area that provincial wildfire hazard mapping shows as swaths of yellow and orange—moderate and high risk—ribboned with red patches of extreme. It's still quite possible to live safely in such fire-prone regions, but it requires an uncommon level of personal responsibility.

Some homeowners took steps to mitigate these risks by fire-smarting their properties in the years before fires arrived, but many didn't. One of the best examples of success is that of local activist and long-time environmentalist Jim Cooperman. After leaving the US to avoid the Vietnam war draft in the 1960s, Cooperman hand-built his multi-storey log home on a hillside overlooking Lee Creek. It is his family's pride. In the years before the 2023 wildfire, having seen the destruction wrought in neighbouring towns, Cooperman hired contractors to aggressively thin out the "beautiful" forest on the hillside behind his home. They cut down 60% of the trees, and stripped and chipped much of the undergrowth, leaving only small stands of pines where once it had been as dense as any forest in BC.

When the Adams Lake fire arrived, it raced down the hillside towards his and hundreds of other homes above Shuswap Lake. But at his neighbours, the fire ripped through the forest and on into their houses, levelling many of them. At Cooperman's property, it reached the broad swath he'd logged and dropped from the tree canopy to the ground. It crawled right up to the rock retaining wall behind the timber-framed patio stage, where Cooperman hosts an annual music festival, and stopped cold. A photo that was published in newspapers across the country at the time shows Cooperman's house illuminated by the orange glow of the fire surrounding it, unscathed.

In the aftermath of a bad wildfire, the trail of destruction can sometimes appear random, or even malicious. A famous photo from the Lāhainā fire shows block after block of homes burned down to blackened foundations, while one seaside house appears to have miraculously survived. How vindictive, it seems, must the fire have been, choosing to spare some homes but obliterating others? The same August weekend that the North Shuswap burned, a devastating fire exploded out of McDougall Creek in the mountains above West Kelowna, about one hundred kilometres to the south. In a matter of hours, it destroyed 191 homes and launched firebrands more than 2.5 kilometres away across Lake Okanagan. The same seemingly random pattern of destruction appeared, with some homes burned to ashes while neighbouring homes were left almost untouched. But often it's not random at all.

Like a lot of people, when I first imagined how a wildfire would hit a city, I pictured walls of flame rolling through backyards and along boulevards like a giant flame-thrower, swallowing houses whole. I imagined ranks of firefighters bravely facing down the fire with hoses, lined up like an ancient Roman shield wall. I could not have been more wrong.

After the McDougall Creek fire, the non-profit wildfire research firm FP Innovations examined what exactly happened and why. They studied thirty-eight structures in total, including twenty-one that were damaged or were lost entirely. What they found is that the destruction of homes in West Kelowna, Traders Cove and elsewhere wasn't really from the wildfire. It was from a wildfire-induced urban conflagration, and the pattern of its destruction was anything but random. "Fire entered the communities almost exclusively via burning embers, rather than direct flame contact, and quickly shifted from being a wildfire to a set of several urban fires," the report said. "Fire spread across neighbourhoods when burning structures produced additional embers, through structure-to-structure ignition between closely spaced homes, and by surface fires spreading across fuels in residents' yards."

In other words, once it hit the city, the fire started burning independent of its forest fire origins. Burning homes, not burning trees, set their neighbours' homes alight. The report describes the main features of homes that were lost, including most of what I'd imagined during the North Vancouver training: "the presence of cedars, junipers, and untreated coniferous trees within 10 metres of a structure, being located on a steep slope, having combustible siding and decks, and an abundance of other easily ignitable material (e.g., firewood, scrap lumber, vehicles, ATVs, fuel cans, recreational equipment) near the structure."

The exact method of ignition was nearly identical across most of the destroyed homes. First, embers landed nearby and ignited something near the home—decorative shrubs, unkempt grass, a pile of firewood—it often didn't matter. Next, the heat from this spot fire would cause the house's siding to catch fire, which in turn often melted the aluminum soffits common in residential construction. Once the soffits were burned away, embers had free access into the homes' attics, where they did the worst of their damage. Once roof trusses and insulation caught fire, it was only a matter of time before the roofs caved in, allowing the fire to evade gypsum walls and other internal safety features meant to contain a regular house fire. From then on, it was only minutes before the doomed house was gone. As it collapsed, it would spew more embers into its neighbours' yards, igniting more spot fires and destroying more homes.

While the fire was still in the forest, before it reached homeowners' driveways, it was at times releasing energy of thirty thousand kilowatts per metre along the fire's front. In some places, it exceeded one hundred thousand kilowatts per metre, far beyond the top of the six-point scale that describes most wildfire behaviour. Firefighters can generally only face down anything less than two thousand kilowatts per metre. At levels beyond two thousand, it's too dangerous to put crews on the ground. At five thousands, even water dropped from the skies is unlikely to have much of an impact. At McDougall Creek, there was enough convective force to easily launch embers clear across the lake.

But the FP Innovations study found something surprising: despite the extreme fire behaviour experienced in the nearby forest, the homes they studied that survived the fire all shared common and relatively simple characteristics: recently mowed lawns with grass less than ten centimetres long, decks in good condition and free of items such as patio chairs, no combustible material within 1.5 metres of the house itself, and leafy, deciduous trees within ten metres of the house. Even in the face of a monster throwing fireballs 2.5 kilometres across the sky, these simple steps predicted a home's chances of survival better than almost anything else.

The FP Innovations study was not the first to highlight how flammable material near homes poses by far the biggest risk during a wildfire. British Columbia's FireSmart policies have for decades encouraged homeowners in fire-prone regions to deal with these ignition risks well before a fire threatens. Yet we continue failing to learn these simple lessons. Frustrated firefighters I've interviewed often described their last-minute and sometimes futile attempts to prepare evacuated homes for the ember storms they know are coming, racing against the clock to remove firewood stacked against houses, decorative cedar hedges, and beautiful tall pines shading backyard gardens. Even those screened-in patio tents homeowners use to escape mosquitos and black flies become giant flammable ember nets that can turn a patio into an inferno in moments.

The FP Innovations study revealed another key finding about the McDougall Creek fire as well, one that goes to the heart of how communities should prepare to face a wildfire but often don't. In the face of an extreme fire, building even giant, last-minute fire breaks on the edges of a city—like Yellowknife did—may well be futile when the forest beyond them is still loaded with fuel. Embers can leap those hardened defences with ease. For fires of that size and severity, there is *no stopping it in its tracks*. To combat this threat, much wider efforts across the whole landscape are needed.

MATTY CAPTAN SHOVES THE BARREL OF A PROPANE-POWERED tiger torch deep into a pile of snow-soaked branches, willing them to burn. It's late December 2024, almost a year and a half since the Adams Lake fire devastated the North Shuswap. As Captan works outside Lee Creek, the occasional snowflake drifts through the forest around him. It seems like an odd time of year to be fighting wildfires, but that's essentially what Captan and his colleagues are doing—albeit slowly.

Captan was working for a local contracting company hired to complete a sixty-hectare wildfire fuel mitigation project in the mountains above Lee Creek and Scotch Creek, communities that now understand better than most how important such projects are.

The work Captan and his colleagues were doing was part of a two-year project to first mechanically thin the forest by cutting down some of the trees, and limbing the rest—cutting off their lower branches and either burning or wood-chipping the leftovers. Fuel mitigation treatments like this one, which take place in the fall, winter and early spring outside of wildfire season, can help prevent the worst of a wildfire's damage, but BC is woefully far behind given that fully half of its forests need the work. The work around Lee Creek should have been done years before the fire, not years after. If it had, the outcome might have been different.

By the time I visited Captan at work on the slopes above the community in December 2024, I'd seen many times over the damage that severe wildfire seasons can wreak, and how much our firefighting systems struggle to contain them when fires explode. I wanted to understand what more we can do outside of fire season itself to protect communities, so I called wildfire ecologist Bob Gray.

Gray has spent a career designing treatment projects like the one Captan was working on for communities across British Columbia. As he explained to me, fuel mitigation means using a number of different methods to reduce the amount of forest material—dried needles and leaf litter, shrubs and understory plants, fallen logs, low-hanging branches and trees themselves—that is available to burn when a wildfire arrives. It's not a question of whether fire-starved forests will burn, Gray told me, it's a question of when and how much fuel they have available when they do.

Nearly one hundred years of aggressively suppressing wildfires has left many Western Canadian forests unnaturally loaded with fuel: vegetation that, under the right conditions, makes any fire that starts far harder to suppress. The amount of available fuel in a forest is referred to as the "buildup index," or BUI. Anything above ninety on the scale is considered extremely hazardous, meaning a fire burning through those densely packed trees will be next to impossible to stop. In August 2023, firefighters trying to contain the Adams Lake fire saw BUIs of more than 130. Firefighters in the Northwest Territories that summer saw BUIs of 200.

Fuel loads like these alone would be enough to seriously threaten many communities in fire-prone regions across Western North America, but the impacts of climate change have supercharged the problem. As the climate continues to warm, it's leading to longer and more intense wildfire seasons. Already-plentiful fuels dry out under years-long droughts like the one preceding the 2023 fire season. The number of extremely hot, dry days increases, and when those fuel-loaded forests finally catch light, they burn with much higher intensity.

The goal of fuel mitigation projects is to help reduce that intensity, bit by bit by changing how a fire burns. The less hot a fire burns, the less of the available fuel it consumes, reducing its destructive impacts. When a fire approaching a town has to burn through a fuel-treated area, it slows down. The severity and temperature drops, it throws fewer embers, generates less of its own wind. Good fuel treatments can take a running crown fire and force it to the forest floor. It won't stop the fire in its tracks, but it can significantly improve the chances that more structures in its path will survive the blaze.

Fuel mitigation work requires careful planning and a unique prescription written for each plot of land to be treated. It's best done in the fall and spring, when cool temperatures and high humidity make the chances of starting an accidental wildfire very low. Sometimes fuel mitigation means selectively logging an area to remove dead, dangerous or fire-prone trees and thinning out the forest. Other times it means setting a prescribed fire to burn away surface fuels and understory shrubs, as Gray himself often does in partnership with Indigenous communities, working with their Elders who know the history of fire on their landscapes. Often it means a combination of both.

The amount of forest that needs to be treated is mind-blowing. A June 2023 report by the BC Forest Practices Board estimated that more than 390,000 square kilometres—nearly half the entire province—has become so loaded with fuel that it's at high or extremely high risk of a wildfire. Another 28% of the province is at moderate risk. Eventually all that fuel will need to burn, but British Columbia, which is considered a national leader in the field of fuel treatment, is decades behind on this critical work. The rest of Canada has barely even begun.

In the past six years, BC's Crown Land Wildfire Risk Reduction program has treated just over 144 square kilometres of high-risk forests around the province's towns and cities, an average of 24 square kilometres per year. By contrast, the State of New Jersey—which is around fifty times smaller than BC—treated 86 square kilometres of land in 2023 alone.

Responsibility for carrying out these costly and time-consuming treatments falls primarily on local municipal governments themselves. The work can cost anywhere from $5,000 to $10,000 per hectare, topping $30,000 for particularly challenging terrain. It's also not very sexy. Politicians tend not to pose for PR photos celebrating fuel mitigation projects as often as they do for new fire trucks or police cruisers. Even with help from provincial or territorial governments, city and town councils often have to navigate thickets of red tape to get even small treatment projects approved and paid for.

At only sixty hectares, the project Captan was working on outside Lee Creek was expected to take about two years from start to finish. Addressing the millions of hectares that require fuel treatments across British Columbia and the rest of Canada is a hugely daunting prospect. Just as the NWT government chose not to follow its own reports' advice because wildfire emergencies are so seemingly "infrequent," it's all too easy for local governments to kick the fuel treatment can down the road. We'll get to it eventually, we tell ourselves, and then we never do.

But while $30,000 per hectare sounds like a lot, the costs of procrastinating on wildfire protections like fuel treatments are far higher. In a bad year, fire-prone provinces like BC and Alberta now routinely spend well over half a billion dollars just on the costs of fighting fires directly. According to a 2025 study by the Canadian Forest Service, the actual price tag of a wildfire can be anywhere from one and a half to twenty times the direct cost of fire suppression alone after accounting for indirect costs like destroyed property, societal disruption, lost economic opportunities, destroyed agriculture land and timber value and more. That means the 2023 wildfire season alone likely cost Canadian taxpayers tens of billions of dollars. By that measure, fuel treatments seem like a bargain.

Along with helping protect a town directly by reducing fire behaviour and limiting embers, fuel treatments serve another critical function as well: They can make a fire fightable. Instead of facing walls of flames and running crown fire like crews did in the North Shuswap, by forcing a fire to the forest floor, fuel-treated areas can make it safe for firefighters to stay in place at the front line for longer and keep up the fight around homes and neighbourhoods. We don't need to surround our cities with moats of concrete or miles of bare rock. Unlike the heroics of last-minute firefighting, fuel mitigation treatments are much less exciting. But they can be much more effective, even if they are painstaking and take years to properly plan and execute. Even once that's done, they require monitoring and maintenance over time. The forest will, after all, grow back. But when it's done well, fuel mitigation treatments around communities can mean the difference between losing homes or saving them. And when a monster fire arrived on Jasper's doorstep in July 2024, they would play a key—if overlooked—role in the town's battle to save itself from the flames.

CHAPTER 8

CALL FOR AID

A hunter strolling north through the Athabasca Valley in 1906, a year before the founding of Jasper National Park, would see the towering, snow-choked peak of Pyramid Mountain dominating the landscape ahead of him. As he quietly stalked elk or mule deer, he would pass beneath the now-famous slopes of Mount Edith Cavell to his west and Kerkeslin to his east, Windy Castle receding into the distance behind him. The walk would be an easy one, following the Athabasca River through a mosaic of open grassy fields and meadows dotted with clusters of fir and spruce. The hunter would pick his way among stands of trembling aspen, using clean sightlines to spot his prey at a distance. From there, beyond Whistlers Peak and Signal Mountain, the landscape spreads out into a wide three-valley gap running east-west below Pyramid's imposing height, punctuated with rolling hills, rocky promontories and a long, arcing plateau where the town of Jasper sits today. The Athabasca River then swings northeast, eventually winding its way past more craggy peaks before flowing out into Alberta's western plains where it joins the mighty Mackenzie River and, finally, the Arctic Ocean beyond.

The mountains the hunter would have walked among are some of the most famous of Canada's Rockies, immortalized on postcards and captured in tourist photos for more than a century. To many Canadians, they are instantly recognizable as Jasper National Park. Without these landmarks to rely on, however, the hunter—were he to make his trek today—would likely not recognize the valley that the Dane-zaa, Aseniwuche Winewak, As'in'î'wa'chî Ni'yaw, Nêhiyawak, Anishinaabe, Secwépemc, and Stoney Nakoda people have called home for thousands of years.

Lori Daniels has been studying these mountain forests for nearly three decades. A professor and researcher at the University of British Columbia, Daniels made her name by using the growth rings of trees to trace the history of wildfire across much of Canada's Western landscapes. The Jack pine and Douglas fir forests of Jasper National Park have evolved not just to survive wildfires, but to rely on them. When a low-severity fire burns through, it consumes dead branches, brush, and smaller saplings, even burning away the lower branches of many trees (what firefighters call ladder fuels), effectively pushing the canopy higher off the forest floor. But the mature trees—especially the firs—often survive, protected from the flames by their dense, insulating bark. When this happens, the fire leaves a scar in the tree's growth rings, and Daniels helped pioneer a method of cataloguing these rings to trace the history of fire on a landscape over the course of centuries.

In Jasper, Daniels's research showed that a hundred years ago these forests would see low- to medium-severity burns every fifteen to forty years. These fires resulted naturally from lightning strikes and, for centuries, from the valley's Indigenous inhabitants who also used fire to carefully thin the forest, promote forage crops and protect their communities. Some of the most frequent low-severity scarring happened around established Métis and First

Nations homesteads, where fire was used frequently to clear and improve the landscape. "There is a lot of fire scar activity right up to about 1900 just before the park was created," she said. After that, evidence of healthy fires on the landscape virtually disappeared.

When Jasper National Park was founded in 1907, the Canadian government forced its Indigenous peoples from their homes, sometimes at gunpoint, removing a critical source of the park's healthy fires. The park, the government decided, was to be preserved and "protected" from all hazards, and for the next hundred years, the fire that its forests relied on was largely eliminated. Firefighting crews perfected the art of pouncing on and stamping out fires whenever they started, denying the forests their key mechanism for thinning and renewal.

By 2002, instead of a healthy mosaic that could be navigated with ease, the Athabasca Valley south of Jasper was choked with wall-to-wall conifers so thick it was impossible to see for more than seven or eight metres in any direction. Around that time, a devastating mountain pine beetle infestation began. A healthy forest might have withstood the onslaught, with its meadows and wetlands acting as natural barriers to the invasive pests' spread. But in Jasper and across Canada's Rockies, the beetles chewed through the densely packed trees with a voracious appetite. Over the next two decades they would transform huge swaths of the park from a carpet of dense green to blankets of rust-coloured red. By July 2024, large tracts of the park had become a standing graveyard of dead grey trees covering the mountainsides.

The town of Jasper sits at the confluence of three valleys with three highway routes in and out. From the southeast, Highway 93—better known as the Icefields Parkway—follows the Athabasca River as it flows north from the Columbia Icefield, hooking right at Jasper and heading northeast out of the mountains and into the rest of Alberta. Highway 16 follows the river in the other direction south from Hinton to Jasper before veering west to follow the Miette River through the Yellowhead Pass towards the BC border, and the town of Valemount beyond.

LANDON SHEPHERD HAS BEEN A WILDLAND FIREFIGHTER AND A fire behaviour expert for more than three decades. July 22 was supposed to be his day off, but at 6 p.m., he was still in the office finishing paperwork. As an incident commander for Parks Canada, his command team was up next in rotation. The weather in the park had been unusually hot and dry for almost two weeks straight, and Shepherd couldn't shake the feeling that a fire was coming. There had already been a couple of smaller ones that summer that his Parks Canada wildland crew had faced down and contained, and with another two-week duty stretch looming, he decided to take advantage of a few hours respite from the heat for a swim with his wife at Jasper's idyllic Annette Lake beach. But as they drove towards the lake, Shepherd started hearing sirens racing down the nearby highway. As they reached the beach, he checked his phone and saw no messages. Maybe it wasn't a fire, he tried to convince himself, but in his gut he already knew.

Shepherd had been anticipating this moment, hoping it wouldn't arrive. In a normal summer, Jasper might experience two or three days of truly extreme fire weather, but by July 22 the park had seen twelve days in a row of record-breaking heat and withering low humidity, the kind of weather that makes pine needles squeak and crunch underfoot like dry snow. In those conditions, everything in the park was potential fuel, and Shepherd knew it better than most. He checked his messages. Still nothing. Shepherd sighed then plunged into Annette Lake's chilly waters. The five minutes he spent swimming would be his last calm moments for weeks to come. Suddenly, as Shepherd swam, a helicopter thundered over the trees towing a Bambi Bucket beneath its belly and sprinkling a fine rain across the lake's calm waters. *That's it*, Shepherd thought. There was no question now. He swam quickly to shore and checked his phone again. Sure enough, a fire had been reported burning around the base of a power pole in the trees beside Highway 16, near the Jasper town dump, about nine kilometres northeast of the townsite. It would become known as the "north fire."

Quickly Shepherd towelled off and started sending text messages to co-workers. *How bad is it? Should I come in?* he asked. His thumb had barely left the send button when a reply popped back. *Yes*, was all it said. Shepherd ran to his car. His wife drove as he sent more messages and started making calls. His first went to Gord Glover, the regional fire coordinator in Edson, Alberta. Though he couldn't even see the fire yet, Shepherd asked Glover if there were any air tankers he could send. The conditions were bad enough that he knew they'd be a welcome help. Glover didn't know what was available but promised to check and get an answer quickly.

Shepherd's next call was to Parks Canada's national duty officer. While Shepherd lives full-time in Jasper, members of his incident management team are based at other Parks Canada locations across the country. They operate like BC's IMTs and often handle emergencies beyond wildfires. In years past, Shepherd's team had helped manage an anthrax outbreak at Wood Buffalo National Park in North Central Alberta, and a hurricane that devastated parts of Prince Edward Island in 2021. When on duty, they are on call and expected to deploy within forty-eight hours. Until then, Parks Canada locations like Jasper are expected to manage any emergency themselves with the resources they have on hand, to allow the IMTs time to arrive and get set up. In addition to the Jasper Fire Department's twenty-eight structural firefighters, the park had three small teams of professional initial attack wildland firefighters who, during the summer months, are on duty to respond at a moment's notice. They also had two teams of Type 2 firefighters, part-time crews made up of regular staff that Parks Canada uses to bolster its firefighting ranks, folks who usually work on trail-building or maintenance crews that get pressed into service as firefighters when needed. The park did not have any of the large twenty-person unit crews that BC and Alberta have available all summer.

Shepherd's IMT duty rotation was not scheduled to start until the following day. But Shepherd told the duty officer he'd need them immediately. "They should start flying tonight, they shouldn't wait till tomorrow," he said.

His last call was to Jasper National Park's superintendent Alan Fehr, asking him to meet at the Parks Canada compound in town to start preparing for what he knew was going to be a significant firefight.

As Shepherd and his wife crossed the Moberly Bridge and hit Highway 16, the north fire's smoke column finally came into view. It was worse than Shepherd had imagined. The column was huge and already billowing with dark black smoke: a sure sign that heavy timber was burning. "We're not stopping that fire," Shepherd thought to himself, but he hoped if they got aircraft on it quickly with some luck on their side, they might be able to hold it to one side of the highway. If they couldn't, the fire might cut off one of the town's three escape routes.

While Shepherd was driving back to town, Parks Canada's wildland fire crews had started to arrive on scene at the north fire. But in the mere minutes it took them to get there, the fire had spread into the nearby trees. The utility pole made helicopter bucket drops too dangerous. The crews did what they could, but the fire was growing fast. Without helicopter support, the wildland crews were outgunned. They radioed for support from Jasper's municipal fire department.

By the time Jasper Fire Chief Matthew Conte arrived with

a crew about a half-hour later, the fire was already starting to crown. The helicopter Shepherd had seen overhead now had plenty of safe targets available, but the water drops hardly slowed the fire at all.

As both wildland and structural firefighters worked to contain the north fire, American tourists Delia Mo and Enoch Leisure were driving south down the Athabasca River Valley along the Icefields Parkway. The pair were approaching the turnoff to the Kerkeslin Mountain Campground when lightning split the sky around them. First one strike, then another. They kept driving, and five minutes later they passed the spot where the lightning had struck. In the five minutes it took to reach it, the fire had already begun displaying rank 5 behaviour, torching individual trees and launching flames twenty metres into the air. "This road is going to be closed within the next hour," Leisure said, as they passed. He could feel the fire's heat on the inside of the car's windows.

Lightning fires often smoulder for hours, days or even weeks before the conditions align to fan them to life. Sometimes those conditions never arise at all, and the fire eventually goes out. For a fire to go from ignition to torching whole trees in under five minutes is almost unheard of. But conditions in the park were so hot and dry that day that all three fires were born into a near perfect pyrogenic environment. The temperature reached nearly 34 degrees Celsius with relative humidity at 18%, classic crossover conditions, exacerbated by wind gusts that topped eighty-seven kilometres per hour.

When Shepherd learned about the lightning fires, he knew immediately that things were grim and getting worse. With the north fire now almost sure to cut off Highway 16 and the south fires showing explosive behaviour, two of Jasper's three escape routes were at risk less than half an hour after the fires began. He consulted with Conte quickly, but there was little ambiguity about what was happening. They were going to need reinforcements.

AT A LITTLE AFTER 8 P.M. THAT EVENING, COLTON BOUTIN WAS looking forward to a late dinner with his wife at the Overlander Mountain Lodge. The picturesque log chalet is nestled in the mountains outside Hinton, Alberta, three kilometres from the gates of Jasper National Park, and about forty kilometres from the town itself. From the dining room, which frequently hosts wedding receptions, family reunions and conferences, big pane windows offer a scenic view down the Athabasca River Valley all the way to Bedson Ridge and Coronach Mountain's usually snowy cap beyond.

As Boutin and his wife were ushered to their table, an eerie silence fell across the bustling dining room. Guests—almost in unison—began pulling out their cellphones. Boutin flicked a glance at the news alerts on his own phone just as the restaurant suddenly lit up with chatter. "Oh my God, they're closing the highway," said a woman at one table. "Oh my God," echoed a man seated at another.

As a senior firefighter with the Hinton volunteer fire department, Boutin knew better than most in the restaurant that night what this news meant. He'd fought almost every big urban interface wildfire in Alberta since 2011, when a wildfire tore through Slave Lake, destroying nearly four hundred structures, including the town hall, the library, and two churches. In 2016, Boutin wore swimming goggles commandeered from a local Walmart to keep the acrid smoke out of his eyes while fighting to defend the Beacon Hill neighbourhood as Fort McMurray burned. In 2023, he fought the early May wildfires in nearby Yellowhead County that ravaged the farming hamlets of Wildwood and Shining Bank, kicking off the most destructive wildfire season in Canadian history.

Instinctively, Boutin knew a fire barely one hour old and already bad enough to close the highway wasn't a good sign. He turned to his wife. "You know, we might get called to this," he said. "This seems to be getting kind of big." After a quick discussion, they decided to bail on dinner and head back to Hinton, just in case. On the short drive back, Boutin started calling and texting other members of the Hinton volunteer department to see if they'd heard anything and within minutes a plan began to form. Boutin stopped by his house briefly to grab his go-bag then headed directly back to the Hinton fire hall to begin prepping trucks and equipment, all before any official call for aid came in.

In the flurry of activity, Boutin continued trading messages with his colleagues, everyone on tenterhooks for the official green light to go. Most of Hinton's on-duty firefighters had been dealing with a car crash on the highway all evening, and they were just returning to their station as news of the Jasper fire spread.

When a town or city calls for help from neighbouring fire departments, it's not always as simple as "Hey, can you come? We need help." Municipal fire departments, especially in small or rural communities, have to run themselves like small businesses. With a population of less than ten thousand people, Hinton's tax base barely covers the cost of its fire hall, a half-dozen trucks, and a few paid staff to administer everything. If the call comes from the municipality of Jasper directly, mutual aid agreements stipulate that Hinton has to cover its own costs to respond—with the understanding that Jasper would do the same. This works efficiently for small incidents like individual house fires, car crashes or other small-scale emergencies. A wildfire battle is a much bigger, and potentially costlier, endeavour. In such a case, requests for aid are usually made in concert with the Province of Alberta, which agrees to foot the bill. But that requires paperwork and bureaucratic sign-off, all of which take time to arrange and, in the meantime, firefighters who are desperate to get in the fight and help their neighbours are often forced to sit on their hands, waiting for an okay to go.

Boutin arrived at the fire hall roughly an hour after he left the

Overlander. Neither he nor his chief had yet received an official call from the province to respond. But, as soon as he stepped inside the empty fire hall, Boutin heard a phone ringing on the main reception line. It was Jasper's Fire Chief Matt Conte and he needed help—fast.

ROUGHLY NINETY MINUTES AFTER THE SOUTH FIRES STARTED, at around the time Boutin started mobilizing the Hinton crews, Shepherd was "eyes on," flying above the south fires in a helicopter assessing the situation. Weather monitoring equipment had detected three separate lightning strikes at 7:05, 7:06 and 7:08 p.m. Staring down through the helicopter's plexiglass window at around 8:45 p.m., Shepherd couldn't believe what he was seeing. Fire from the first strike had grown to about fifty hectares already, about the size of a soccer field. The second was five times as big and the third was even larger, with whole groups of trees torching together, well into rank 5 territory. Some armchair observers would later question why the south fires didn't get hit with water bombers from the start. Before lifting off from the helipad at the Parks compound, Shepherd had gotten a call back from Glover, in Edson. It was bad news. The nearest air tankers were tied up fighting fires elsewhere in Alberta, and others available from BC were grounded by heavy smoke at their airports. Firefighters in Jasper would have to make do with the few helicopters they had on hand.

Even if aircraft had been available, large tankers often struggle in steep, mountainous terrain, and there were no viable water sources big enough in Jasper for smaller skimmer planes to refill at, meaning they'd have been able to make only a single small drop before departing. And even then, the aircraft wouldn't have helped. If Shepherd had had whole squadrons of heavy-lift Blackhawk helicopters with 3,700-litre buckets already airborne the moment the fires had started, it would still have amounted to spitting on a campfire. All three south fires quickly merged into one three-thousand-hectare monster, and was only getting started.

Meanwhile, he and Conte were already stretched thin dealing with the north fire which, burning much closer to town at the time, seemed to pose the bigger threat. Transport trucks were jackknifing on the highway trying to avoid driving through it. The park that week was full to bursting with tourists, and along with trying to manage the fire, crews were also racing to the Snaring Campground and the park's overflow camping area northeast of the townsite, trying to get hundreds of families and other vacationers out of harm's way.

At 8:28 p.m., the municipality issued the emergency alert Boutin had seen on his phone at the Overlander, and Parks Canada closed both the Icefields Parkway south and Highway 16 northeast towards Hinton. "This is a preventative measure due to the number of fires," the town's statement said. "There is no immediate threat to the town of Jasper."

As he flew over the fire, Shepherd worried about how long that would remain true. He took stock of what natural features he had, planning a tactical defence in his head. Between the fire and the townsite there was the river itself and Highway 93A, which runs east-west below the town. There was also the Miette River, which flows into the Athabasca at the edge of town, plus several rocky ridges and wetlands, natural barriers that crews could reinforce with fire guard and ignitions to help slow the fire down. There was also Highway 16, the TMX oil pipeline, and the CN railway line. Each of them was wider and more robust than dozer guard would be, and they were already in place. It was more defensive lines than an incident commander could reasonably hope for, perfect positions for crews to fight and burn off from—if conditions allowed.

But those options were being quickly wiped off the board. With the south fires growing so explosively, Parks put its focus on tactical evacuations of the thousands of campers along the Icefields Parkway at the Wabasso, Whirlpool and Kerkeslin Campgrounds, all of which lay in the growing fire's path. These evacuations were carried out by Parks nonemergency staff. They were mostly twenty-something students working summer jobs as trail guides or maintenance crew workers, but they went into the campgrounds without hesitation, trying to coordinate thousands of scared families and get them out safely, all while an enormous smoke column reared over them. Less than two hours after they were emptied of tourists, both Whirlpool and Kerkeslin were completely burned over.

Unlike Yellowknife, Parks Canada and the municipality of Jasper had spent significant time and effort mapping the wildfire risks around the town. They had used wildfire modelling to identify risk areas and created set boundaries—which firefighters call trigger points—around the town. If a fire breached these trigger points, evacuation orders wouldn't be a political decision, they would be automatic. As Shepherd flew over the fires in the fading evening light, there was no doubt the trigger points had been reached. The town issued a full evacuation order at 9:59 that night.

BRETT IRELAND GREW UP IN JASPER, PART OF ONE OF THE OLDest families in town. His uncle Richard became the town's first—and so far only—mayor in 2001 when Jasper became a formal municipality. In 2005, he founded the Jasper Brewing Co. with two other local boys. And though he eventually moved to Fort McMurray, opening a brewery there and others in Banff, Edmonton and Calgary, he's often back in his beloved town for work.

On Monday, July 22, he was staying with his parents in Jasper, leading staff training as his team prepared to open another Jasper business, a distillery called The Maligne Range. Wednesday was to be the new distillery's opening day, and Ireland and his staff were working diligently to get the final touches in place.

The business is only a few blocks from his parent's house, so Ireland had ridden to work on his bicycle, surprised at how much dust was kicked up on the streets around town. Something eerie hung in the bone-dry air, and Ireland wasn't used to it.

Staff were only a couple hours into the training when word of the first fire started to spread. Many of them were young, like staff tends to be in many tourist towns across Canada. Some had their own vehicles, but many didn't. Ireland remembers one couple frantically trying to buy tenant insurance. Ireland made sure each of his staff members had a way out of town, then he jumped back on his bicycle. As he rode back to his parents' house, he saw gridlock already stretching through town. Tourists queued up for gas, already fearing the worst.

He arrived back at his parents' place just as the first evacuation alert went out. Ireland was lucky. As the fire danger ratings had climbed higher over the preceding twelve days, his parents had taken precaution and prepared to leave. They even had the negatives of their precious family photos safely packed and ready to go. Lots of other families were not as prepared. Traffic on Jasper's streets was backing up even worse as the minutes ticked by. He helped his parents load their small RV with supplies and then they waited, hoping for the traffic to die down.

Part of the gridlock was due to Parks Canada staff blocking

the exits from the townsite to the highway so they could keep it clear for all the campers fleeing the campgrounds in tactical evacuations. The last thing anybody needed was panicking motorists crashing into each other and blocking the only remaining safe route out of town.

By the time the official evacuation order was issued just before 10 p.m., the whole town was bumper-to-bumper traffic. With nowhere they could drive, Ireland's family decided to get a few hours' sleep, taking turns to wake up and check the traffic. Finally, at around 1:30 a.m. Tuesday morning, they managed to get on the road. They joined the convoy of vehicles headed west through the Yellowhead Pass. The drive from Jasper to Valemount, BC, is only 120 kilometres and it normally takes less than ninety minutes. There were so many people trying to leave Jasper at once that it was just after 6 a.m. by the time Ireland and his family pulled off the highway into Valemount, exhausted but safe.

JUST AS THE EVACUATION ORDER WENT OUT AT 10 P.M., COLTON Boutin and his crews were rolling down the highway towards Jasper. Passing the now-closed National Park gates, they could see the north fire was growing rapidly. As they approached the town, fire was burning on both sides of the road now, right up to the ditches. They had to drive the final few kilometres from the Jasper Park Lodge to the centre of downtown on the left-hand side of the road past the same gridlocked traffic that Ireland and his family were trying to wait out.

By the time they arrived, almost everyone else was already on the roads trying to get out. After assigning two of his crew to patrol for spot fires, most of Boutin's contingent checked into a hastily emptied hotel. "There's keys on the front desk and towels in the bin," the clerk said, adding that they were welcome to whichever rooms they needed, but none had been cleaned yet: All the staff had evacuated along with the guests. The clerk was on her way out the door too. Boutin and his crew weren't about to complain; they fell into restless sleep atop dirty beds. They would start preparing Jasper's defence at first light.

CHAPTER 9

WE'RE NOT GOING HOME TODAY

Regan* craned her neck to peer over her colleague's shoulder and out the helicopter's windscreen. Down below, the chopper's shadow traced along the Athabasca river, flickering over treetops and the rocky shoreline. Up ahead was a smoke column from a fire Regan could barely believe was less than a day old. It was only 9:30 a.m., and already the fire was roaring in the forest crown.

---

* Regan is not this firefighter's real name. They asked to be identified only with a pseudonym because Parks Canada—like most wildland fire agencies—does not allow its firefighters to speak freely to the press, and doing so without authorization could jeopardize not only their career, but the careers of their crew mates as well.

On the flight from Banff that morning, the Parks Canada wildland crew was relaxed, cracking jokes over the helicopter's intercom headsets, excited to be sent to a fire. But when they rounded a valley corner and saw the fire itself, chatter on their headsets went quiet.

"Yep," Regan thought to herself. "This is pretty serious and we're not very far from town."

Within minutes, Regan's crew landed and went straight to the Parks Canada compound. Municipal crews had been setting up structure protection around the town for hours, but there was still much more to be done to prepare for the coming fire, and the scene was busy with fire trucks zipping around the town. Jasper's wildland crews had fought the fires the best they could until roughly 3 that morning, and now they were back again after only a few hours' sleep. Regan couldn't see much of an organized command structure yet. Her crew waited for hours at the compound, anxious to be assigned a mission.

By Tuesday morning, firefighters on the north fire were seeing some improvements. Along with the handful of helicopters Parks Canada had in Jasper already, Alberta Wildfire had sent others, and bucketing on the north fire now appeared to be slowing it down, at least a little. On the southern front, however, things had deteriorated badly. Fire behaviour in the Athabasca Valley was so extreme that ground crews couldn't attack it directly. Instead, they focused on helping the Jasper and Hinton firefighters deploy structure protection sprinklers, and last-minute fire-smarting around residents' homes.

There was one cause for hope: Air tankers were finally available, and command sent them to attack the south fire, hoping that retardant drops might slow the fire's advance.

Flying air tankers in mountainous valleys like Jasper is dangerous at the best of times. Wind shear off the mountains can create unstable flying conditions. In order to effectively hit their targets, tanker pilots have to get unnervingly close to the ground. To help with this, they follow much smaller spotter aircraft called "bird dogs." These bird dogs will fly the target ahead of a tanker, laying a strip of smoke through the air behind them along the best flight path. As they pass, they broadcast a loud warbling sound from speakers pointed at the ground, warning anyone in the tankers' path that a drop is imminent. Then the tankers swoop in low behind, lining up the target with the smoke to guide them. Then the pilots open the tankers' belly doors, unleashing up to nine tonnes of water or bright red fire retardant that fall in giant billowing plumes.

The pilots flying in Jasper on July 23 faced almost impossible conditions. The fire was burning so fiercely it generated its own weather, with wind speeds estimated to have reached more than two hundred kilometres per hour. The turbulence was so bad, the pilots' heads banged against their cockpit walls as they fought to hold their planes steady long enough to drop. Two air tanker groups reached Jasper that day. As the first group lined up for their bombing run, crosswinds swirling in the valley tossed them around like their planes were made of paper. They managed to make one drop, with each plane releasing fire retardant in long strips along the flanks of the fire as they roared over the treetops. But instead of falling to the ground and hitting their targets, the bright red retardant was sucked horizontally into the fire on hurricane-force winds, rendering it useless.

The second group tried to circle around and come at the fire from behind, flying up the Athabasca Valley from the south, but it was no use. The winds had gotten too strong. The air attack officer told Shepherd he'd never experienced anything that bad in twenty-five years of flying on fires. If the winds died down they could come back, but for now they had to retreat.

ON TUESDAY MORNING, BOUTIN AND THE HINTON CREW WOKE early and got straight to work. Fire behaviour modelling that day predicted the south fire could hit the town on Friday. That meant crews had a little more than three days to prepare. Boutin's crew met up with Jasper deputy fire chief Don Smith and his firefighters, working hand-in-glove to set up sprinkler lines and other defences around the town. They put sprinklers on the hospital, the fire hall and the water treatment plant, all the critical infrastructure they would need to protect if the town had any hope of survival. Fire modelling at the time showed the fire was expected to reach town by Friday at the earliest. Boutin's crew worked late into the evening, then headed back to Hinton to get a good night's sleep. In the morning, they returned at first light and continued helping prepare the town's defence alongside their Jasper colleagues.

The Cabin Creek neighbourhood sits at the far southwest end of town, separated from the rest of Jasper by a small strip of green space full of trees. As the rest of town arcs towards the north, it's protected by the width of the Athabasca River's frigid waters. Cabin Creek has only the much narrower Miette River separating it from the slopes of Whistlers Peak. It was an almost unbroken stretch of forest full of fuel, making Cabin Creek one of the areas most likely to take a frontal assault from the fire as it burned towards the town.

As effective as sprinkler and water cannon systems can be, they—like all defences—have their own weaknesses. For sprinkler systems, one of the main ones is time. If they are turned on too early, they can run out of the fuel that runs the pumps, or water or both. But if they're turned on too late, they won't have enough time to fully soak the surrounding area, limiting their effectiveness. Timing is everything.

People who live in major cities tend to view water as an essentially infinite resource: Turn on the tap and it's always there. But the mechanics, logistics and simple physics required to make that happen are complex. Well-resourced fire departments will plan out the defence of a city or a town like military leaders laying battle plans against an approaching army. They decide well ahead of time where to put strategic sprinkler lines, which hydrants to hook up to, which neighbourhoods will need the most protection. Such detailed plans for an entire town can take months and sometimes even years to complete, and they rely on one single all-important resource: water pressure.

Most municipalities plan their water delivery systems to accomplish two main tasks: Deliver water to people's homes so they can shower, shave or do the dishes, and have enough on standby through fire hydrants for crews to effectively battle structure fires. A major five-alarm house fire might have a dozen fire trucks responding, pouring thousands of litres on the burning structure for hours, all without any noticeable drop in water pressure across the system. For almost a century across North America, this was sufficient.

But municipal water systems are largely not designed to defend against twenty-first-century megafires. In January 2025, residents of Los Angeles learned this lesson the hard way: When those three fires sparked simultaneously under extraordinarily dangerous conditions, it wasn't long before hydrants in the Pacific Palisades and other higher-elevation neighbourhoods began to run dry.

At first, this seemed like a preposterous error on the part of fire officials and the city: How could a metropolis so accustomed to wildfires simply run out of water? But the reality is much more complex. First, despite hysterical early reports, firefighters in LA didn't run entirely out of water. The majority of municipal water systems rely on gravity to generate the pressures needed to push water out of thousands of taps, shower heads, garden hoses and fire hydrants. When the system is functioning normally, pumphouses and reserve tanks positioned throughout the system can resupply the main reservoirs faster than the water flowing from all those outlets. But when you turn on every tap at once, the systems often struggle to keep up. As hundreds of hydrants are opened at the same time, the pressure begins to drop, and the highest-elevation hydrants start to fail first.

In Los Angeles, this was compounded by some lack of communication between the various municipalities' public works departments and the fire departments. In at least one case, a key reserve tank in the Pacific Palisades neighbourhood was down for maintenance, further weakening the system's overall capacity, and firefighters weren't warned ahead of time. But the reality is that even a region as accustomed to wildfires as Los Angeles County doesn't have the infrastructure to fight fires of the speed and ferocity seen that January. Even if the system had been at 100% efficiency, it wouldn't have been enough to stop the fires. In Jasper, access to water would also prove a turning point in the fight to save the town.

EVENTUALLY REGAN'S FOUR-PERSON CREW GOT THEIR MARCHing orders. They were split into teams of two, with each team assigned a "twenty-pack" of part-time Type 2 firefighters to oversee. These were the kind of regular staff—such as maintenance crews or folks who typically work on trail-building—that Parks presses into service when a wildfire threatens. As soon as the teams were organized, they got to work doing last-minute firesmarting of homeowners' properties—moving anything flammable as far away from houses as possible, making sure windows and doors were sealed up. Regan and her crew spent hours going house to house, where they often chucked things like patio furniture, sports equipment and propane bottles out into the street or dragged them to the ends of driveways. They started in Cabin Creek and worked their way east.

While these kinds of last-minute efforts around a home can appear like arranging deck chairs on the *Titanic*, they can make a meaningful difference to a home's chances of surviving an ember storm. Regan and her crew were happy to do the work, but seeing just how much of it there was to be done left a clear impression on her. "People do not understand what fire-smarting is," she thought, as she lugged another wooden patio chair into a driveway. Some houses had cans of gasoline stored in the backyard; others had large mesh bug-screen tents, dead patio plants, decorative door mats and more.

As they worked, Regan noticed some of the firefighters who lived in Jasper appearing to struggle. "I saw a lot of people just kind of spiralling," Regan said. "We were going from house to house and they're like 'Oh my god, this is my hockey coach's house, I hope it's going to be okay.'" Being from Banff gave Regan an advantage: This wasn't her town. That wasn't her coach's house. She couldn't imagine trying to keep emotions in check and do the job if it were.

They continued prepping structures until about 6 p.m. when their supervisor ordered them to take a break for the evening. While the intent was to manage fatigue—firefighters were preparing for a large-scale battle in the morning—Regan and her crew were eager to keep working. She figured they had maybe seventy-two hours to finish protecting the town, and she wanted to get it done.

To burn off her anxious energy, Regan and a colleague went for a walk after dinner. Around 9 p.m., they reached a ridge overlooking the area. They could see the fire south of town, the glowing flames in the fading evening light. "This fire is not three days away," she thought to herself. Later that evening, there was a knock on Regan's hotel room door. A group of Jasper residents had gathered outside and they were scared. Unlike the rest of the town who had already evacuated, this group had been designated as essential workers and asked to stay. Lots of folks were in that position—doctors, paramedics, emergency dispatchers—people who'd never faced a wildfire before.

One of them asked if it was safe to stay in the town that night, even though they'd been told to. Regan felt a pang of sympathy. She gave one of them her phone number, saying "I'll text you if anything changes, but you're safe for now," she said, wondering how long it would take before she had to follow through on her promise.

THE NEXT MORNING REGAN'S CREW WAS BACK DOING MORE last-minute structure prep. The south fire continued burning so hot it was still well beyond what firefighters could face head-on. Along with Parks' wildland crews, a twenty-person unit crew from Alberta had now also arrived. As they finished prepping the last structures in town, Regan felt the winds shift. They radioed back to command asking for another assignment, but without one immediately available, they were told to stand by.

Meanwhile, the fire south of town was gathering strength. Within hours of ignition, all three south fires had merged into one 3,500-hectare monster. As it tore up the valley towards Jasper, it passed below the famous Marmot Basin ski resort, where something began to unfold so vanishingly rare it had only been documented once before in Canadian wildfire history. The fire started to spin.

Exactly how this happened is still not entirely known. As of October 2025, scientists from the Northern Tornado Project were still investigating, but some experts believe that as the Jasper wildfire grew, it generated extreme heat that rose thousands of metres into the sky, punching through the clouds. The indraft caused by this rising mass sucked more air down Jasper's three intersecting valleys. When the air currents collided, the theory goes, they formed a cyclone of unimaginable proportions, further feeding the monster and creating a towering column of swirling fire and ash nearly half a kilometre wide and reaching to the heavens.

On the ground, it would have looked like Armageddon. Fire-generated winds reached 260 kilometres per hour, as strong as the worst tornados in the world. Winds were so extreme they blasted rocks into tree trunks like bullets, ripped steel campfire rings from the ground, and threw a 450-kilogram shipping container into the Athabasca River. They flayed the bark from whole mountainsides of trees, ripping them from the ground roots and all, long before the fire itself even reached them. As the fire burned, it scattered thousands of tree trunks in huge arcs—like a giant's game of pick-up sticks—and left them so disfigured that forest ecologists would later struggle to identify them. The heat was so intense it shattered boulders and incinerated the soil. The smoke and ash coming off the fire didn't look so much like a column or plume as it did like a giant wall, filling the Athabasca Valley from peak to peak, a flaming fire front burning nearly eight kilometres wide.

Sitting on a curb by the police station as they waited, Regan and her fellow wildland colleagues did their best to project an air of calm and confidence. Around them were a group of Type 2 firefighters who had helped them prep the homes around town. Some had never seen a wildfire up close before. Unlike professional firefighters like Regan and her crew, Type 2 firefighters are often only given a basic wildland firefighting and fire entrapment avoidance courses. Having done both of those courses myself, I can say with confidence they don't truly prepare you to fight a wildfire for the first time. They give you just enough information to recognize the risks and to follow the instructions of the more experienced veterans you will (hopefully) have around you. Like their essential worker colleagues that Regan met the night before, the Type 2 crew members had been given a choice: They could stay to face the fire or they could choose to evacuate with everyone else. Incredibly, only a small handful chose to go.

As they waited for their next assignment, a buzz began to move through the group. An email had arrived from incident command: The Parks Canada maintenance yard, located in Jasper's industrial area near the river, would be everyone's fallback position. They were to shelter in place there if things went totally sideways and escape from the town became impossible. As news of the plan spread, a few Type 2 crew members hovered tentatively nearby, screwing up the nerve to come over and talk with Regan. Eventually, one of them asked what *shelter in place* meant. "Are we going to have to use those tinfoil fire shelter things?" Regan recalls one of them asking in a quavering voice. "Because I don't have one."

The moment left Regan rattled. As a veteran firefighter with almost a decade of experience, she knew as well as any that Canadian firefighters don't use emergency fire shelters. They're not supposed to need them, yet here was this rookie, asking Regan if she was being asked to risk dying in a burnover to save the town, and she had chosen to stay anyway.

"Wow, these people are brave," Regan thought to herself. "They're just blindly following us, doing whatever assignment we ask, and they don't even know what to do when this fire actually hits town." But they were what the town had, and so they would try.

AT AROUND 3 P.M. ON WEDNESDAY, JULY 24, JASPER'S DEPUTY Fire Chief Don Smith was standing on the front pad of the town's fire hall, watching as the smoke column from wildfire south of town filled the sky above him. As fire crews raced to finish bolstering the town's defences, Smith privately hoped that maybe they'd get lucky. Maybe the fire would snake its way along the far side of the valley with the river between them and slither away to the east. But the winds started to shift. "It was starting to make these runs across the valleys, up the hills," Smith told me when I interviewed him months later. "It was just building and building, getting bigger and bigger."

On the west side of town, below the Cabin Creek neighbourhood, Boutin and his crew watched it too. They were working next to the river, where a series of wells pump water up into the town's reservoir, high on the mountainside to the north. Some protection was already in place, but Boutin helped add more, just to be safe. The water wells would have to be defended at virtually any cost. Without them, there'd be no way to resupply the reservoir and keep water flowing to the town's fire hydrants, sprinkler lines, and the dozens of fire trucks that were arriving to help from places like nearby Parkland County. As Boutin and his crew worked, the fire's smoke column loomed higher and darker. He knew what it meant. He turned to his crew: "Guys, we're not going home tonight. This thing is gonna be here today."

Thirty minutes later, Jasper's air raid siren sounded, calling everyone back to the fire hall.

CHAPTER 10

THE BATTLE FOR JASPER

WEDNESDAY, JULY 24, 2024, 4 P.M.

It's safe to say that most people are, if not afraid of wildfires, at least distinctly uncomfortable around them. Ben Bartlett is not one of those people. A thirteen-year veteran of Alberta Wildfire's initial attack program, Bartlett is more comfortable repelling out of a hovering helicopter into a wildfire than he is sitting through a staff meeting. For most of those thirteen years, Bartlett was a rap-attack firefighter, one of a small number of elite initial attack crews sent to battle the most remote or inaccessible wildfires in the province.

In 2019, when the Alberta government cancelled the rap-attack program Bartlett loved, he was gutted. Rather than uproot his life in Hinton, he took a job at the local mill and a volunteer position with the Hinton municipal fire department, leaving the world of wildland firefighting for good. Or so he thought.

On the afternoon of Wednesday, July 24, Bartlett enjoying a rare day off from that mill job when his phone buzzed with an all-hands call from the fire department. He headed straight for the Hinton fire hall. Ash had formed a thick blanket across his truck and the sky overhead was boiling with smoke. "Hinton was black," Bartlett said. "It was like a volcano went off."

Racing towards Jasper on an empty highway, Bartlett and the other Hinton firefighters drove through the north fire. Trees torched beside the highway. Yet as they approached the town, the smoke suddenly gave way to a cobalt-blue sky and sunshine. It was like driving into the eye of the storm. Behind them they could see the north fire's smoke column rising over the mountain ramparts and blowing towards Hinton. To their south they could see a wall of smoke bearing down. The strangely calm moment left Bartlett unsettled.

By around 3:30 p.m., Boutin and his crew were just about finished setting up sprinkler lines in the Cabin Creek neighbourhood when the Jasper air raid siren sounded, calling them back to the fire hall, even though some of their sprinkler lines were not yet complete. "It's coming," command told them. "We think it's a couple hours out, maybe an hour." Joining Jasper's twenty-eight firefighters and Boutin's Hinton crews, firefighters from nearby Parkland County had also arrived, bringing the total structural firefighting force to around forty people. There were also hundreds of wildland firefighters from Parks Canada, Alberta and BC.

It was time to switch over from preparation to active defence—a sort of man-the-battlements moment. That meant reorganizing their trucks from a standard structural firefighting stance to what's called a bump-and-roll tactic, like the one I'd seen firefighters practice in North Vancouver's mountainous neighbourhoods. When operating efficiently, a well-trained crew can sometimes bump-and-roll along a street, knocking down spot fires without even stopping the truck.

As the smoke column loomed over Jasper, incident command was forced to make a tough call. As the situation deteriorated, Shepherd and Christine Nadon, the town's emergency planner, were on a video call with Jasper Mayor Richard Ireland, updating him on the situation. Suddenly, Shepherd stood and left the room. When he came back a few minutes later, he spoke to Nadon directly.

"Christine, we have to leave," Shepherd said.

She was surprised. "Right now?" she asked.

"Yes, right now," Shepherd replied, glancing out the window at the fire's column looming on the Whistlers Peak slopes above town.

It would be almost impossible for incident command to maintain operational control in the middle of the firestorm that was about to arrive. It was safer, and smarter, to move the whole command structure down the highway to Hinton while they still had time. But incident management teams are cumbersome beasts, with tentacles that touch operations, logistics, planning and finance, and uprooting one like Jasper's and moving it an hour down the highway in the middle of an urban firestorm would be no easy feat. There were sure to be communication lapses.

As the fire got closer, Jasper Fire Chief Matthew Conte stayed mostly at the fire hall, trying to keep track of his firefighters and oversee the battle in concert with his deputy, Smith, who was out in the field helping dispatch crews. As everyone prepared to move the command structure to Hinton, they all began packing their equipment: laptops, radios, everything they'd need to continue managing Jasper's defence from eighty kilometres away. Everyone, that is, except Conte. No matter what else happened, he and his firefighters were not going anywhere.

There was another problem. As the fire raced down the valley towards town, there was no way firefighters could hope to face that kind of ferocity on the ground, and Shepherd and the rest of incident command knew it. He watched from town as the fire ate its way closer, blowing past every potential defensive position he'd mapped in his head two days before and upending his understanding of what a wildfire is really capable of. Like many firefighters in the fire's path, Shepherd had grown up in Jasper. He knew the park and its terrain intimately, and to think of it all burning was almost more than he could bear. While he'd seen plenty of big angry fires in three decades of firefighting, he wasn't used to seeing his home in a wildfire's crosshairs.

With no way to fight the fire on the ground, and air tanker drops now proving effectively useless, there was only one tool the command team had left in their arsenal that had any hope of success: aerial ignitions. But unlike more traditional aerial ignitions, the goal wouldn't be to simply burn up the forest fuel between the fire and the town. These ignitions needed to act like their own severe wildfires, generating enough heat and convective winds to interrupt the firestorm's swirling ferocity. They needed to somehow split the main fire in two. But for that plan to work, the ignitions would have to be carried out flawlessly under extreme conditions.

The plan called for two ignitions, one running up the slopes of Whistlers Peak southwest of town and another up Signal Mountain directly across the valley from Whistlers. The two mountains stand like guardians protecting the entrance to the town. The ignition chopper lifted off around 4:40 p.m. and began its runs up Whistlers' flanks, its PSD machine dropping ping-pong balls of fire as it went. With its payload delivered, now they just needed to do the same thing eight kilometres across the valley. Fighting the clock, they managed to get several strips of fire laid down along the slopes of Signal Mountain before winds and smoke overtook them.

Regan was working with her crew to prep some of the last houses of their assignment when she heard the ignition pilot's frustrated calls come over her radio. "It's too damn smoky," she recalls him saying as they were forced to land, not knowing whether the Hail Mary had worked.

As they prepared to face the fire itself, Boutin heard the same calls, worried that the fire was now much closer than he'd thought. In their setup, Boutin's crew had tied one end of the water well sprinklers to a hydrant beside the road and the other to a pump placed down in a ravine that fed directly from Cabin Creek itself: a belt-and-suspenders approach that would guarantee the sprinklers had maximum water pressure. But hoping to preserve precious water and fuel, Boutin had left them switched off until the last minute. He thought they'd have more time.

While attempting aerial ignitions under such extreme conditions had been a risky call, the ignitions specialist, who had arrived from the Yukon only that morning, was considered one of Canada's best. Firefighters in Jasper also had another significant advantage. Starting in 2002, the Park—and specifically former Park Warden Alan Westhaver—had begun studying ways to reduce wildfire risk with fuel mitigation treatments. In wide swaths around the town, they had cut, slashed and burned strategic sections of the forest to reduce fuel loads and, most critically, the likelihood of those areas becoming crown fires. Today, Westhaver is widely considered to be the godfather of Canada's FireSmart program, which relies on methods he pioneered in Jasper. Without those fuel treatments to rely on, the aerial ignitions so close to town would likely not have been possible.

But possible and successful are two different things. For a single ignition helicopter with so much ground to cover in so little time, the odds were long. Like the air tanker pilots the day before, the ignition helicopter pilot faced an eight-kilometer-wide fire front generating hurricane-force winds as he swept the sides of Whistlers Peak and Signal Mountain with fire. The fire was advancing so fast he had barely half an hour of viable flight time. In that short window he managed to make eight runs before being forced back to base. Whatever fire he'd managed to lay down across the town's mountain gates would have to be enough.

At first the burns appeared to have little impact on the behemoth burning towards Jasper, but gradually the tide began to turn. As fire raced up Whistlers Peak, it generated enough of its own internal heat to influence the winds ripping through the valley. Slowly, almost imperceptibly at first, the south fire began to pull itself apart. As the second ignition up Signal Mountain got going, the tug-of-war on the main fire increased. Winds flowing west down the Miette valley from BC helped even more, pushing the bulk of the fire east, parallel to the town instead of directly onto it.

As the fire grew up Whistlers' slopes, it finally merged with the ignition lines and picked up speed. The faster and hotter it burned, the higher its convection drove the plume of smoke, ash and embers into the sky over Jasper. But when the fire reached the mountain's peak, it hit the rocks of the barren, treeless summit and ran out of fuel. Without the heat needed to sustain its enormous height, the column collapsed, falling like a giant tower crashing onto the town. Winds that had been sucking the fire up the mountain and into the stratosphere now reversed and flowed back down into the valley. They blasted the entire southwestern edge of Jasper, bathing homes, businesses—everything—in embers, ash and smoke. Day turned to night. The fire had arrived.

Boutin and his crew raced to get the sprinklers turned on. They reached the hydrant end at the roadside first and opened it, ensuring at least some water would flow, but they still needed to reach the other end down the hill, tied into the creek.

The winds howled. Opening a truck's door risked it getting ripped clean off. Firefighters had to wrestle to keep their helmets from flying away in the gale. Boutin's crew drove right into the teeth of the storm, trying to reach the water well sprinkler line.

The bottom end of the line was connected to a wildland fire pump: a simple two-stroke engine that requires a pull cord to start. It sat as close to its water source as possible, with only a short intake hose feeding it from the creek. Parking near the hydrant, Boutin and another firefighter jumped out of their truck and scrambled down the ravine to the pump. They had time for one good yank on the pull cord before Boutin heard a voice come over the radio. It was a structure protection specialist on a rise overlooking the area, and he could see they were about to be overrun. His voice crackled frantically over Boutin's radio: "Get the hell out of there! It's gonna go right over you!"

Boutin and his partner were forced to flee, abandoning the pump's engine without managing to start it. With at least the hydrant feeding the line sprinkler line, they knew the system could function at 60%, maybe even 70% capacity. It would have to do. They ran back up the ravine and jumped in their truck. Just as the engine revved to life, embers ignited around them. They pumped the gas, trying to get ahead of the fire so they could attack it. But the wind was too strong. The fire was spreading too quickly.

Forced into a tactical retreat, Boutin and his crew raced back to the Highway 16 junction—just in time to see the historic Maligne Lodge in flames. "Holy shit, the hotel's on fire," he exclaimed. Behind him, and deeper into town, six or seven new columns of thick black smoke boiled into the sky. Houses were burning down. Maybe an hour had passed since the fire had breached their defences. To Boutin, it felt like ten minutes.

Across town, all hell was breaking loose. When the ember

storm swept into town, it ignited spot fires everywhere. Those fires in turn ignited more spot fires. Firefighters' organization completely broke down. Divisions collapsed; strike teams merged into each other. Everyone scrambled to fight fires that seemed to be sprouting from the very earth itself.

Back at the fire hall, Smith, Jasper's deputy fire chief, was dispatching his crews as fast as he could. Once the rest of command left Jasper, Conte and Smith shifted into a sort of roving patrol mode, driving laps around the town to check on fire crews and call in more structure fires. One of the first buildings to burn was Wicked Cup, a coffee shop attached to the Maligne Lodge. Then the Petro-Canada gas station exploded, then Home Hardware caught fire, then a hostel. And then more houses.

THE JASPER WILDFIRE HAD A RAGING FLAME FRONT AND thousand-degree heat. But if it weren't for the vicious, hurricane-force winds, all the defensive features Shepherd had identified, along with the lines of carefully placed sprinkler systems, might have stopped the fire from tearing north up the Athabasca Valley. But it wasn't to be. The fire devoured the final eight kilometres of closely packed Douglas fir and lodgepole pines in mere hours, leaping from tree to tree with ease. Whenever the raging flames struck a barrier, the wind helped spur an air assault, throwing millions of embers—some the size of pine needles, others the size of iPhones—into backyards, onto roofs, and into needle-clogged rain gutters far behind the front line. Everywhere firefighters looked, there were flames.

At about the same time Smith and Conte were frantically dispatching crews to burning homes around the town, Bartlett and the other reinforcements from Hinton began arriving on scene. They beelined to the now empty fire hall. Bartlett walked up to the fire hall's second floor, looking for whomever was in command. Upstairs, he stopped to gaze out the big pane windows, looking west from the fire hall towards the Cabin Creek neighbourhood and Whistlers Peak beyond. There were homes on fire only a block or two away. For all his wildland firefighting experience, Bartlett had never seen anything like this. "Holy . . . ," he breathed quietly, processing the view of a town on fire. He was looking at a full-blown urban conflagration.

Meanwhile across town, Regan began to wonder whether they too should prepare to evacuate, just in case. Wildland firefighters aren't trained to battle burning buildings, and they lacked the kind of specialized, self-contained breathing apparatus that structural firefighters use routinely. She and two colleagues raced back to the apartment where they were staying to gather their belongings. It was just after 5:30 p.m. when they arrived at the apartment. They walked in just as the windstorm struck, massive gusts rattling the windows and bending the trees outside. Her crew leader's voice came over the radio.

"The column's collapsed, where are you?" he asked, sounding anxious. He told them to muster at the Parks compound to resupply their gear. Outside the sky was darkening. Regan and her crew mates raced to the compound. The facility sits in a large gravel yard and includes several large industrial buildings. It also includes a lot of parked staff cars and a rail yard next door full of creosote-soaked replacement railway ties. By the time Regan's crew arrived, those ties were on fire, throwing embers everywhere and belching clouds of acrid black smoke.

They started filling piss packs—backpack-mounted water tanks with a hose and a hand-operated pump—getting ready to fight. They also planned to hook a portable water trailer to their pickup truck. As they struggled to get the water tank connected, other Type 2 firefighters started arriving at the compound, just as command had directed them to. One of them approached Regan, telling her that he worked in the wildlife division and all of their ammunition was stored in a nearby locked shipping container. He pointed to it: "Maybe don't stand there."

As Regan looked around, it dawned on her that the Parks compound wasn't as safe as command had hoped. It had only one road in and out, and no alternative escape route. The railway ties were already on fire, and there was plenty left there to burn. Her crew finished filling their piss packs and hooking up the water trailer. Then, they tried to drive back into town. The smoke was so thick they couldn't see where they were going.

With all the fire trucks and other vehicles roaming around Jasper, and with little in the way of cohesive coordination, the last thing they needed was a head-on traffic collision. Gingerly they reversed the truck back into the Parks yard and waited. The sound of exploding gas stations and propane tanks echoed over the howling wind. From what Regan could see, the town was now completely surrounded by fire; all their escape routes were compromised.

Regan had worked wildfire for years and been in her share of close scrapes, but sitting in the Parks yard, with toxic smoke billowing from the burning rail ties and no clear escape route, she felt a lump rise in her throat. "Shit, I might not make it out of this one," she thought to herself. She considered calling her parents and saying one last goodbye.

Across the town, chaos was unfolding. Other wildland crews were being told to find whatever "achievable objectives" they could wherever they could. With the IMT on its way to Hinton, and Conte and Smith trying their best to assign crews directly, tactical command had broken down. Jasper was in a fight for its very survival, and Regan was stuck sitting in a pickup truck surrounded by smoke so thick she couldn't see, unable to do anything but listen to her wildland colleagues calling to each other on the radio.

"Hey, start the pump!"

"Strangle that hose line!"

"I need more water pressure!"

In many places around town, the other wildland crews were fighting backed up against carefully fuel-treated areas. As the Jasper blaze burned across the hillsides towards places like the Edith Lake neighbourhood, it was raging at rank 6, torching whole stands of trees at once, and throwing flames sixty or eighty metres into the air. But as it approached the neighbourhood, it ran into the areas where homeowners, using the guidelines Westhaver helped develop, had thinned the forest and helped reduce the fuel loads. As the fire burned through, it ran out of the type of ladder fuels it needs to reach a forests' upper branches. This allowed firefighters to stay longer, equipped with hoses and lines of sprinklers, facing ground fire instead of crown fire.

As Regan's crew sat in their truck listening to the radio chatter from other crews, the frustration of being unable to help started to mount. Regan's crew leader lobbied to go back into town and do what they could, but conditions in the yard itself were worsening by the minute. They had no idea if things would be any better among the burning houses.

After some debate, the crew decided to go back. They crawled along the road towards town in their pickup, the smoke so thick their headlights barely cut through the gloom. Up ahead, trees had been blown down across the road. Their only way out of the yard—and the only route in for anyone trying to find shelter—was blocked. She and her crew leader got out of their truck and grabbed their chainsaws. The smoke outside the truck was still so thick, she struggled to see. At this point, she assumed escape from the town itself was impossible. Fighting down her fears, Regan focused on what was right in front of her.

"Okay, I know how to start a saw," she said to herself, squeezing the chainsaw's throttle. "I know how to cut a tree and move it, and that's all I have to do right now." Once the roadway was clear, they piled back into the trucks and headed for town. A few minutes later, the skies above them suddenly opened, showing a brief glimpse of blue. Regan felt a brief see-saw of hope. Elsewhere other crews—both wildland and structural firefighters—were battling the blaze. One of the unit crews from Alberta Wildfire was working on a ridgeline above town, with a good view of the unfolding battle. Their radio chatter was encouraging.

"Okay, people are safe," Regan thought. "It's not so socked in. We can work with this." She breathed a sigh of relief—then she noticed all the burning houses. Fire raced from rooftop to rooftop. What she saw wasn't a forest fire burning inside a town. It was the town burning itself down.

At first it wasn't clear what—if anything—Regan's crew could do to help. Columns of inky black smoke billowed into the sky from more than a dozen burning buildings. By now the Maligne Lodge hotel was gone. An apartment complex was in flames. Front yard trees were torching, covering whole blocks with embers. Her crew kept driving, looking for achievable objectives. At the corner of Ash and Patricia Streets, they found one.

Fire was burning from the eastern edge of the Cabin Creek neighbourhood, out of a copse of trees and into tinder-dry grass. Beyond the grass was a row of houses. Several had large decadent conifers in their backyards. If the fire reached the trees, they would shower the block with embers, and the houses would likely be lost.

"Grab the torch," the crew leader said, as they all jumped out of the truck. A sense of determined calm came over Regan—*this is something we can save*, she thought as they set to work.

The two other firefighters grabbed their piss packs from the truck and started wet-lining around the houses. Regan and her crew leader worked laying lines of burning torch fuel carefully into the grass, letting the wind carry the flames forward. Anytime they built too much intensity, the crew would knock back the flames with jets of water. They worked methodically, burning off strip after strip of grass until they had a wide patch of nice, clean black surrounding the houses. It was textbook wildland firefighting, and it worked.

Their spirits lifted by the small win, the crew jumped back in their trucks and went looking for another target, their adrenaline pumping. They chased spot fires around town for the next two hours, knocking down what they could.

AS THE FIRE BLEW INTO JASPER, CONTE COULD SEE THEY WERE facing a potentially town-erasing event. With the incident management team now safely in Hinton, it was time for more people to get out. Regan's earlier hunch was right: The fire had become too dangerous for the wildland firefighters to stay. At roughly 8 p.m., command ordered them all to leave.

When the call to evacuate came over the radio, it was like a weight lifted off Regan's chest. "Wow, we're actually going to make it out of here," she said.

By 8:15 p.m., all the wildland trucks were organized into a convoy on the highway headed to Hinton, but it still wasn't clear the road through the north fire was passable. In one of the last vehicles to leave town, Regan's crew mates had their radio's volume turned up high, listening for reports from the front of the convoy about whether the road was drivable. When they finally got word that it was, Regan felt both relief and a pang of guilt for leaving her structural fire colleagues behind. She wished them a silent *good luck* as she and the other wildland firefighters reluctantly headed out of the smoke. Jasper's fate now rested on the volunteer structural firefighters alone.

SOMEONE—BARTLETT DOESN'T REMEMBER WHO—GAVE THE order to protect the hospital, so "we jumped in our trucks and literally drove around the corner, and everything was burning around us." A truck parked in front of the hospital was already on fire. Bartlett and his crew knocked that fire down and started patrolling around the hospital, looking for spot fires. In planning the town's defence, the Jasper fire department had identified several pieces of critical infrastructure to save at all costs. If they lost the water treatment plant, it could be years before residents would be able to return to whatever else was left. The same was true of the fire hall and the hospital—if they lost those, there would be almost no point in trying to save anything else. No one can live in a town with no water treatment plant. A national park that hosts 2.5 million visitors a year can't function without a hospital to care for them. As Bartlett and his crew patrolled around the hospital, he couldn't help but feel overwhelmed. A part of him wished he was back on a wildfire somewhere in the backcountry, far from burning trucks and exploding propane cylinders.

While most people risk getting lost as soon as they step off a marked hiking trail, wildland firefighters like Bartlett are experts at navigating the backcountry. But one of Bartlett's biggest challenges that night in Jasper was quick navigation. It isn't a particularly big town—banana shaped, not quite six hundred metres across at its widest point, and only about 3.5 kilometres end to end. But it's not Bartlett's town, and without reliable paper maps on hand, he kept getting turned around. To make matters worse, the amount of smoke and ash blowing through Jasper that evening reduced visibility to mere metres. "You just couldn't get out of it," he later told me.

On a wildfire, crews don't tend to spend that much time in heavy smoke. Getting caught in heavy smoke means you're upwind of an advancing flame front and might even be at the head of the fire. In other words: You're probably in the wrong place. If that happens, firefighters will usually try to reposition so that they can get out of such a precarious position. In Jasper, everywhere felt like both the wrong place and the necessary place to be. Even for someone as experienced with fire as Bartlett, being in the middle of an inferno like that was bewildering. At the end of the day, a wildfire is pretty simple. There's the head of the fire, the right flank and the left. You know which way the wind is blowing, so in a sense, it's easy to navigate. Wildland firefighters are always fighting from the outside in. There are helicopters like eyes in the sky supporting you. But instead of working around the edges of the Jasper fire, Bartlett and his colleagues were in the centre of it. "It's hard to describe," he told me months after the fire. "I feel like my jaw was on the ground the whole time."

There were other significant differences too. Compared to what he'd hear on a wildfire, Bartlett thought the radios in Jasper were awfully quiet. Tactical positioning and communication is key on a wildland fire. In ideal scenarios, it's a carefully coordinated ballet of people, aircraft, machines and flames. In Jasper, there just wasn't that much to say: Everything was on fire, and everyone was already doing what they could to put those fires out. The extreme wind speeds—at times topping 110 kilometres per hour—meant there were no aircraft overhead, and no point in trying to describe the details of the fire's movement. It was simply everywhere.

BY LATE EVENING ON WEDNESDAY, HUNDREDS OF STRUCTURAL firefighters from departments across Alberta and BC were heading to Jasper to help. Some would take hours to arrive, some days. But a handful made it in time to help, including crews from Parkland County and others from Grande Cache, a hamlet just outside the Rockies north of Hinton.

As Boutin and the other Hinton firefighters joined with those from Jasper, Smith and Conte did their best as mobile field commanders, patrolling around town while working the radio, dispatching crews where they were needed. But by now, fires were starting around town faster than anyone could report, much less coordinate responses to. In the Cabin Creek neighbourhood, two things made it worse. One: The homes—many of them stately single-family wood-framed houses that wouldn't look out of place on a Christmas card—were packed close together. Jasper has struggled for decades with a housing shortage exacerbated by its restrictive municipal boundaries that prevent building further out into the National Park. Two: Many of those beautiful homes also sported the iconic cedar shake roof shingles that, while offering a charming aesthetic perfectly in keeping with a quaint mountain town, are also highly receptive to embers.

In Jasper, as embers took hold of one picturesque roof in the Cabin Creek neighbourhood, it in turn cast embers onto the neighbouring homes' roofs. Soon, the fire was racing through blocks of homes so fast that firefighters struggled to keep track of what was burning and what wasn't: the urban equivalent of a crown fire. A fire crew would knock down a spot fire on the roof of one house and race off to another, believing they'd saved the home only to return on a patrol lap an hour later and find the house completely annihilated.

Structural firefighters are mostly trained to fight individual building fires and prevent conflagrations in the first place, and they have become very good at it. But as Deputy Chief Smith later told me, that's largely because under normal conditions, even the worst structure fire takes a long time to burn. "If this house across the street caught on fire," Smith says, pointing out through his office window, "it would take hours. It would start off, it would get bigger, it would burn for a good two hours and then it would finally collapse into the basement." That window gives firefighters the time they need to set up engines and pumper trucks, to tie into hydrants, to do whatever they need to suppress the fire and—even if they ultimately can't save the home itself—they can at least keep the fire from spreading.

In Jasper, Smith saw whole apartment complexes reduced to ashes in half that time as flames shooting from ground-floor windows reached halfway across the road.

To understand the science behind this, it helps to imagine a campfire. As night settles over your campsite and the temperature drops, you huddle closer to the fire and push the logs closer together—stoking up the fire. If the fire's burning too hot, you back up and move the logs farther apart to spread out the heat. The same is true of burning houses. Once a conflagration gets going, a block of buildings packed close together act like logs themselves, only you can't simply push them apart with a fire poker. Instead, you often have to simply bulldoze them.

In the same way wildland firefighters will cut miles of dozer guard through a forest, trying to separate the fire from its fuel, firefighters in Jasper quickly went into triage mode. If a row of five houses was already on fire, they wouldn't bother trying to save the first four. Instead, they'd bring in heavy equipment to plow the fifth one down, creating a fire break that would hopefully save the other as-yet-unburned houses on the rest of the street.

As the fire chewed further into Jasper, spreading through Cabin Creek and other western neighbourhoods, Smith and Conte juggled resources all across town, trying to adapt to the fire's chaotic spread. Within a few hours, most of Cabin Creek would be lost. Though it was painful to let the whole neighbourhood go, at this point Conte was more worried about saving what he could. As he, Smith and fire chiefs from some of the other departments started telling their crews to fall back, Conte happened to drive past his own house, and it was on fire. He didn't dispatch anyone to try and save it. He barely paused to look, and continued working the radios, pulling crews back to the parts of town they had at least some hope of saving. He wasn't alone.

Eight members of the Jasper Fire Department lost their homes to the fire, and there was not a single report of anyone breaking focus or breaking down. Grief would have to wait. They all just kept working to save other people's homes, even as their own went up in flames.

AS THE EVENING WORE ON, CONTE COULD SEE THE FIRE WAS making a run towards the downtown core, where three rows of buildings paralleled each other north along Connaught Drive and Patricia Street. Losing the downtown core would rip the heart out of Jasper's tourism economy, crippling it for years, if not decades. But just like lodgepole pines packed tightly together, the downtown was essentially three long strips of continuous fuel. Most of the buildings adjoin each other and are fronted with picturesque post-and-beam wooden facades in keeping with Parks' vision of an iconic mountain village. Despite years of municipal efforts to transition the town to more fire-resistant roofing, some of the downtown still had highly flammable cedar shake roofs. If the fire got into those buildings and took hold, the firefighters could lose not just the downtown, but the whole town.

Conte chose Miette Avenue as their fallback position, a defensive line they'd have to hold to protect the entire north end of town. Miette arcs northeast from the base of Pyramid Bench to the main drag along Connaught Drive. Its two single lanes are separated down the middle by a small boulevard of grass and spruce trees. Directly north of it is the hospital, the fire hall and the Jasper Park Visitor Information Centre. The RCMP station, the Jasper-Yellowhead Museum and Archives, two public schools, the aquatic centre and the town's arena and community centre all lie within its gentle curve.

As the risk to the downtown core continued to grow, Smith decided to take matters into his own hands. What they needed was an aerial truck set-up, but the other firefighters were frantically working all across town and besides, not everyone knew how to run one. Not everyone, maybe, but after nearly forty years of firefighting, Smith himself sure did.

He raced back to the fire hall, took the aerial pumper truck and drove it to the intersection at Hazel and Patricia Streets. He set it up by himself, pumping out an umbrella of water over the four corners of the intersection, hoses snaking this way and that around him to hydrants. Smith stood at the controls raining sheets of water as he swept the nozzle back and forth from three stories high for hours. But while Smith's defence of downtown helped keep new spot fires from igniting, it wasn't enough to contain the already stubborn blazes burning deep inside buildings along the strip. They needed another fire break.

Conte describes bulldozing buildings with characteristic matter-of-factness. On Connaught Drive, a string of businesses was burning—"Fully involved" as Conte says—so heavy equipment was brought in to knock the buildings down with fire engines on standby. At around 9 on a typical July evening, restaurants along Connaught Drive, with their curbside patios overlooking mountains and the river valley, would normally be bustling with tourists enjoying steaks, seafood and pasta as they basked in the warm evening sunset. Now they bulldozed, along with a liquor store, a bike shop and a cannabis dispensary—all sacrificed in a bid to stop the fire's spread.

On the other side of the block, along Patricia Street, the roof of another building was on fire, and the flames were spreading quickly, spewing embers onto nearby buildings, including the Earl's restaurant and its vulnerable cedar shake roof. If the Earl's went up, it would have threatened the town's historic Athabasca Hotel—lovingly referred to as the "Atha-B"—a block from the fire hall itself and a favourite watering hole for Jasper's firefighters. Heavy equipment plowed another fire break into the buildings along Patricia Street, just in time to save it.

IT WAS JUST APPROACHING 9:30 P.M. WHEN BEN BARTLETT FIRST noticed the spot fire on the hospital's roof, just behind their new fallback line at Miette Avenue. He radioed other trucks in the area. Several converged, but they couldn't reach the flames from the ground. Though the building had sprinklers set up at various points on its roof, their protective domes couldn't reach the flames either.

A creature of his wildland training, Bartlett grabbed a ladder from his truck and scrambled onto the roof, waiting neither for direct orders nor for permission. The hospital's roof had been undergoing minor repair work before the fire, and embers had collected in a pile of garbage bags and plywood left there. Bartlett started grabbing burning debris and throwing it over the side of the hospital to the ground below, where the firetrucks' hoses could reach it.

Generally speaking, climbing onto the roof of a burning building in the middle of a firestorm is very much against the rules, and for good reason. First, if the roof is more heavily involved than it appears from the outside, it could collapse, trapping the firefighter and forcing others into a risky and likely futile rescue. Second, if firefighters had clambered onto every burning roof in Jasper that night, they'd never have kept ahead of the blaze. Given the importance of the hospital to the town's survival, Bartlett figured it worth the risk, at least for the two minutes he was up there. When he climbed down, his deputy chief, Colton Boutin, was waiting for him.

"You know, you're not supposed to do that. You're not supposed to go on a roof," Boutin said sternly. Then he grinned. "Good job. You saved the hospital," he said, quickly adding, "but don't do that again."

There was no time to revel in the success: Across the street, kitty-corner to the hospital, a whole block of houses was already gone. Boutin and Bartlett went back to work alongside everyone else. An hour later, they had met up with other firefighters from Jasper, Valemount and Yellowhead County and were working their way street by street through the blocks northwest of the hospital.

At one point, they entered a back alley. On one side, all the homes were starting to burn, and one home's garage was already fully engulfed. On the other side, nothing had caught light yet. The alley itself had become an impromptu fire guard, the kind of defensive line Bartlett was used to fighting from. As they'd been doing all night, Bartlett and his crew fought it like a wildfire, soaking the houses on the green side and knocking down hotspots as soon as they sprang up. They let the homes in the black continue to burn themselves out. Then they moved on.

They worked this way for hours, in alley after alley, street after street, until they reached a mobile home park. Bartlett watched flames consume the homes in minutes, each one adding more embers to the storm. Some firefighters tried to soak the burning trailer homes, attempting to cool them down and reduce the ember cast, but their efforts were futile. Another excavator was brought in to do its awful work, crushing the burning trailers in on themselves. Better to expedite the combustion while crews were nearby to contain it rather than let the trailers continue to burn for hours.

As night wore on, access to water started to become a challenge. Some of the sprinkler lines firefighters had painstakingly set up only hours before now had barely any water coming out at all. The more the fire destroyed, the more it was able to weaken firefighters' remaining defences. Every building in a neighbourhood has water lines running to it from the city's mains. As homes burn down to basements, waterlines become exposed. Each burned home then starts uselessly pumping out thousands of litres per hour. Combine this with dozens of fire trucks, hundreds of sprinkler defence units, and Smith flooding the downtown with the aerial, and it wasn't long before Jasper's reservoir started losing pressure and risked running dry altogether. To combat this shortage, crews in Jasper started strategically shutting off water mains to the neighbourhoods that were already beyond saving. This helped preserve their water supplies, but it also created a knock-on effect. Along with interrupting the flow to destroyed houses, it meant the street hydrants in those neighbourhoods also went dry, limiting where fire crews could hook up to replenish their trucks.

By 11 p.m., the town was nearly full of firefighters. Departments from about forty different towns across Western Alberta and Northeastern BC had answered Conte's calls for aid. Roughly 280 firefighters—and, as Smith put it "god knows how many fire trucks"—were now racing across town, putting out spot fires and trying to save burning buildings.

By about twelve o'clock, the reservoir was down to about 5%, and "we had very, very low water pressure," Smith said. After more than seven hours of sustained firefighting, both the water supply and the firefighters were nearly spent. But miraculously, it seemed so was the fire. As July 24 turned into July 25, the weather finally began to shift. The winds died down, and the waves of embers abated. The fire, it seemed, had finally run out of breath. When Smith finally shut down the aerial truck sometime close to midnight, he made a startling discovery.

It had actually started raining.

CHAPTER 11

RESET

When dawn finally broke over Jasper on July 25, it revealed the terrible cost of a hard-fought victory that, at first, was difficult to see through the lingering haze. The southwestern end of town was a smoking ruin, with dozens of family homes reduced to ash-filled pits. Firefighters would spend days drowning basements. The Cabin Creek neighbourhood, which had taken the brunt of the ember storm, was almost completely gone. The historic Anglican church, which had sat at the corner of Geikie Street and Miette Avenue since 1928, was now just a lonely stone chimney reaching towards a grey and cloud-filled sky. But as heartbreaking as those losses were, it could—and in some ways should—have been far worse. Of the town's 1,113 structures, an incredible 755 were still standing. More than two-thirds of the town, and all of its critical infrastructure, had been saved in a battle against odds so long it's a miracle no one was killed.

Out in the valley, the destruction was even more intense. On the mountainside below Marmot Basin ski resort, one of the most important tourist draws in the park, evidence of the fire's savagery left Shepherd speechless. Sometimes, if a wildfire burns hot enough, it can sterilize the soil, killing the micro-organisms, the seeds, the root structures and the mycelium networks that fire-adapted ecosystems rely on to recover. At Marmot Basin the soil wasn't just sterilized, it was completely gone, burned to ash by thousand-degree heat and then blasted away by hurricane-force winds. Across a huge swath of mountainside where the suspected fire tornado touched down, not a single tree is left standing. It didn't look like a wildfire at all. It looked like the site of a nuclear blast.

Other parts of the park now resembled an alien landscape. In areas around the Edith Lake cabins, where fuel treatments had helped tame the fire's behaviour at least a little, some of the trees appear to have developed leopard spots. This can happen when the temperature in a forest rises so fast that the sap in the trees boils before the tree burns. The boiling sap causes sections of the blackened bark to erupt in little jets of steam, leaving behind round pockets of white.

On the slopes of Signal Mountain and the Maligne Canyon lookout, an even weirder sight remained. When super-heated gasses driven by the fire's ferocious winds ripped through stands of fir and lodgepole pine, they stripped the bark from trees completely and bent them forward almost ninety degrees in smooth, graceful arcs. The extreme temperature differential between the windward and leeward sides of the trees caused their cell fibres to dry out at different rates, locking them in place. Experts call it fire freeze. Whole mountainsides of gleaming white tree trunks now appear frozen in time, cursed to stay bent forever before the fury of a phantom wildfire.

Faced with a fire like this, Jasper had been almost certainly doomed. The fact the town had survived at all was an incredible success, but one that was at first hard for many to recognize.

As the fire ripped into town on Wednesday night, snippets of video and a few photos showing horrifying scenes began rocketing around the internet. They left thousands of people with the impression that the whole town had been destroyed, a perception that persisted for days.

With the wildfire itself still burning beyond the townsite, Parks Canada and the municipality struggled at first to meaningfully rebut this. Early statements by officials at a press conference on Thursday were vague, saying somewhere between one-third and one-half of the town had been lost.

Professional press access could have helped provide crucial context, but journalists couldn't get into town. As officials had done in the North Shuswap,* Parks Canada largely refused to allow journalists into Jasper for nearly a month after the fire. Unlike after previous fires, these restrictions continued even after the evacuation order was lifted and town residents were allowed to return. When asked about this by understandably frustrated journalists, Parks Canada claimed it was protecting the privacy of Jasper at "a time of profound loss, devastation, and grieving."

This bears underscoring. In sixteen years as a journalist, I've never seen a public agency deny reporters access to the site of a major public disaster on the grounds of privacy alone. I've also never seen the news media subject to more restrictions than residents themselves, and the precedent it sets is alarming.

For weeks, the only meaningful press photos to emerge from Jasper were made for the Canadian Press by photojournalist Amber Bracken, who was briefly allowed into the town days after the fire front had moved on. Her visit was as part of a media tour arranged by Alberta Premier Danielle Smith, who was photographed alongside emergency managers looking suitably mournful as she surveyed destroyed homes.

---

* And as is common for virtually every wildfire in Canada.

In the absence of robust independent coverage, misinformation around the Jasper fire swirled into a political storm. In October 2024, parliamentarians questioned everyone they could during hearings in Ottawa, including Conte and Shepherd, slicing the disaster up into their own self-interested angles. The federal Conservative Party was keen to pin blame for the fire on then–Prime Minister Justin Trudeau's Liberal government, accusing it of having mishandled the pine beetle infestation that left so much dead fuel across Jasper's landscape. Others accused Parks Canada of refusing the aid of firefighters who could have helped, including those hired by private insurance companies who, critics claimed, were forced to stand by and watch while the town burned.* Premier Smith complained that her government had been shut out of the fight to save the town, even though Alberta Wildfire crews and aircraft had been critical in the battle's early hours.† Incredibly, hardly any of the critics mentioned the cedar shake roofs, which were a holdover from official Parks Canada policy stretching back decades.

It got so bad that Jasper Mayor Richard Ireland decried the political infighting, telling local media it was exacerbating the pain felt by residents who'd already lost everything. "The present atmosphere of finger-pointing, blaming and both partial and misinformation is, from my perspective, beyond merely an annoying distraction," Ireland said. "It delays healing. It introduces fresh wounds and fosters division, precisely at a time when we need recovery and unity."

---

* According to firefighters who actually fought the Jasper fire, the majority of these private firefighters arrived Thursday morning, too late to have been any real help. At that point, the town was already full of fire crews from nearby communities, most of whom had little left to do. "By Thursday morning we were already drowning basements," Boutin told me.

† A report later commissioned by the Municipality of Jasper found that Smith and her cabinet members' interference during the fire had made things worse, not better. Smith publicly disputed the report, characterizing it as "unfair," and demanding an apology.

Just as they had after the North Shuswap fire and many others before it, public officials in Jasper found themselves fighting a new battle for the narrative of the fire. And despite the hours of testimony and litres of ink spilled, virtually all of this public debate missed the point.

AS HEARTBREAKING AS THE DAMAGE TO JASPER WAS, WHAT happened there was not a failure. The biggest lessons we should take from it aren't about what went wrong. Jasper represents our new reality, one in which losing only a third of a beloved town amid the conditions firefighters faced should be seen and celebrated as an incredible, if implausible, win.

The town had vulnerabilities before the fire. Jasper's fire hydrants used a different thread pattern than virtually every other municipality in BC and Alberta, and the small number of adapters they had on hand were quickly used up. This meant that as hundreds of firefighters poured into town to help, they were unable to connect to the town's main water supply. Thankfully there was a saving grace in the form of a lone purple fire hydrant, sitting just east of the Cabin Creek neighbourhood. Its purple paint denotes non-potable water, in this case because it connected directly to the stream of pipes feeding untreated water from the intake wells by the river to the city's reservoir in the hills above town. This meant that the consistently strong water pressure inside the purple hydrant kept up even as other hydrants across the town began to fail—a virtually unlimited supply.

As fire crews across town fought to keep the flames from spreading, drivers of special tanker trucks called *water tenders* raced back and forth from this one lone purple hydrant, filling their rigs and then resupplying crews directly, allowing the fight to continue.

Jasper's many cedar roofs were another clear weakness, and like the forest conditions themselves, one that developed over decades. Much as they'd have liked to, town council couldn't simply rip off every cedar roof and replace it with metal all at once.

Even with these chinks in its armour, unlike most other Canadian towns, Jasper had been comparatively well-prepared for a fire. Over the course of several years, Conte and others had built a robust wildfire protection plan. Their system of fuel treatment areas around the town was more complete and extensive than any other Canadian city ever hit by a wildfire. They proved invaluable to its defence. Unlike the leadership in Yellowknife, officials in Jasper had practised emergency wildfire evacuation scenarios, including running a tabletop evacuation drill just six weeks before the fires struck. On top of that preparation, they also got extremely lucky. Aside from the cedar shake roofs and chaotic, last-minute fire-smarting around people's homes, almost everything that could have gone right did. If a single dice roll had gone the other way, if the westerly winds down the Miette valley hadn't arrived or if the aerial ignitions had failed, the town could have been easily wiped off the map.

The biggest lesson of all from Jasper isn't political; it's a message about the type of fires now possible in our current climate-heating world. As Lori Daniels's research shows, wildfire has always been a presence in Jasper's valleys. But the severity of the 2024 fire was something new.

When I first saw the burn scar below Marmot Basin, photographing it from above in a helicopter in July 2025, I could not believe what I was looking at. I'd seen the aftermath of high-severity fire before. Hiking through the burn scar of the Elephant Hill wildfire, an enormous blaze that levelled a mobile home park in Boston Flats, near the Village of Ashcroft, in 2017, I saw the ghostly-white ash piles reaching like giant skeletal fingers across the forest floor, all that remained of whole trees completely incinerated. In 2021,

I slipped and slid my way across greasy, hydrophobic soil left behind on hillsides in the aftermath of the White Rock Lake wildfire. That particular fire had torched most of Monte Lake, BC, before running more than forty kilometres, all the way to the shores of Lake Okanagan, where it levelled more houses in the hamlet of Killiney Beach. Witnessing that kind of destruction is always unsettling, but what I saw flying over Marmot Basin left me rattled.

Across an area nearly ten square kilometres, hardly any trees were left standing. Most were torn from the ground and left strewn in smooth, circular rings, their trunks pointing into the fire, not away from it. As Shepherd explained to me, the blowdown pattern indicates they were uprooted by the sheer force of the fire's indraft. Those that weren't ripped clean out of the ground were snapped off two or three metres above the ground. And these were not small trees. "Some of the trees are these three-hundred-year-old Douglas firs that survived the valley burning wall to wall at least twice in the past 250 years," Shepherd told me. The first major fire happened when the trees were young, sometime in the 1700s. Then in 1889, more than half the entire park burned, and still the trees survived.

There are several reasons the same trees weren't so lucky in 2024. Many politicians pointed towards the valley full of dead pine beetle–killed trees—and while politicians misdiagnosed the problem, it's true that those trees contributed to the fire's intensity. In what's called the "red phase" shortly after a beetle outbreak, those dead trees are still covered with their dead, rust-red needles. When a fire takes hold in a forest like this, the fuel load of dead needles in the canopy causes the forest to act like so many matchsticks, packed close together. Light one of them, and the fire races ahead across the match-heads launching galaxies of embers as it goes. It's fast, but it's not as intense as the Jasper fire. Under normal conditions, traditional firefighting methods can still be effective.

The trees in the Athabasca Valley had long since turned grey, littering the forest floor with pine needles, and leaving grey tree trunks and branches standing naked on the hillsides. Over time, branches break off, dead trees fall over. The majority of available fuel is no longer in the forest's canopy, it's on the ground. With the forest canopy now naked, the extra sunlight dries out all that downed fuel even more, making more of it available to burn.

This process can contribute to a fire's residency time—how long it stays burning in one place before moving on in search of more fuel. In normal conditions, a fire's residency time is relatively brief, sometimes only thirty to sixty seconds. But when grey-phase beetle-killed trees are present, the fire has more material on the ground to burn. It stays in place many times longer, burns hotter and—as happened at Marmot Basin—can create the conditions for a fire tornado, bringing the fire back over the same ground again and again, consuming more fuel, feeding itself.

On the six-point fire behaviour classification scale, a fire becomes a rank 6 blaze when its energy output exceeds ten thousand kilowatts per metre. At this point fire suppression is essentially impossible. The McDougall Creek wildfire, which destroyed nearly 200 homes near Kelowna, BC, on the same weekend as the North Shuswap blaze, exhibited fire intensity of 100,000 kilowatts per metre. A Canadian Forest Services study that took detailed plot samples and reconstructed the Jasper fire's behaviour found that, at its height, it was generating more than 386,000 kilowatts of energy per metre, more than thirty-eight times hotter than the top of our existing scales and seven times more energy than the nuclear bomb dropped on the Japanese city of Hiroshima across a flame front more than eight kilometres wide.

But for any of this to happen, the fire first needed the kind of extreme conditions that climate change is making more common. It needed two whole weeks of high heat and frighteningly low humidity to parch even the coarse surface fuels, the ten- and hundred-hour fuels like branches and logs, getting them dry enough to make them, and everything else in the park, available to burn. Next, the fire needed—and got—high winds. And once it ignited under those conditions, no force on heaven or earth could stop it. Fires in these conditions are, as many firefighters put it, "unsuppressible."

"It should set off alarm bells for everyone concerned about the impacts of climate change," Lori Daniels told me.

To have stopped the Jasper fire from reaching the townsite, Parks Canada would have had to clear-cut enormous swaths of a beloved national park—or somehow undo more than a century of wildfire suppression. Jasper, like the rest of the country, is now dealing with the problems that took more than a hundred years to manifest. Overcoming that legacy is going to take hard work, a lot of money, and a lot more time. There are no shortcuts.

FOR VENTURE CAPITALISTS, THE WORLD OF WILDFIRE TECH IS burning bright. The race for the next great wildfire solution is on, and it's garnering huge amounts of capital. Dozens of companies are competing to develop the hot new AI-driven wildfire detection and modelling tools, and more want to marry AI-driven detection with swarms of autonomous drones to attack new wildfire starts, replacing human pilots flying water bucket drops, or to trigger aerial ignitions with airborne autonomous flame-throwers. There is even a company that wants to both detect and suppress wildfires with beams of light alone. Or, as they put it in their own techno-bro babble: "Use intense light beams to repeatedly detonate the atmosphere by ion-based avalanche breakdown and extinguish the fire with the resulting acoustic shockwaves . . ."

Reading these claims, I couldn't help but feel skeptical. I don't know about you, but the idea of *repeatedly detonating the atmo-*

sphere with high-intensity light beams sounds an awful lot like space lasers to me. These "solutions" seemed like magical thinking, but then again I get easily frustrated trying to navigate the new operating system anytime my iPhone updates itself. I am not, what you would call, a technologist.

So, I called former wildland firefighter and current wildfire technology expert Mathieu Bourbonnais. He's an assistant professor at the University of British Columbia Okanagan, and he studies how new technologies can be used to help manage wildfire risks and even improve how we live alongside wildfires. One project he helped develop in partnership with the BC Wildfire Service uses 5G-enabled remote sensors to autodetect wildfire risks on the landscape, allowing faster response times and better resource planning.

Most fire weather modelling relies on regional weather stations that can't provide hyperlocal readings of weather behaviour inside a wildfire zone. Bourbonnais's stations can. They're easy to deploy to wherever the wildfire risk might be growing, recording and transmitting—in near real time—all the relevant wildfire indices like temperature, relative humidity, precipitation, wind speed and even the moisture content of duff layers on the ground.

In describing his research, Bourbonnais pointed to the McDougall Creek wildfire. By luck, weeks before the fire ignited Bourbonnais had deployed his prototype weather stations into the forests around West Kelowna to test them. Fifteen of his prototypes wound up in the fire, and the data they were feeding out—before the fire destroyed them—told Bourbonnais from the start how bad McDougall Creek could be.

Bourbonnais's prototypes told him all sorts of things about what the fire was doing, including the wind speeds inside it and the temperatures it was generating. Everything pointed to the fire's potential to jump nearly three kilometres across Lake Okanagan. He fed all this information to the BC Wildfire Service in real time, giving them a significant new advantage. Having this kind of geographically specific vital information could help save lives by speeding up evacuation orders and helping firefighters better understand exactly where the risks are at any given moment.

Our conversation turned to the severity of recent fire seasons, and what technology might be used. Bourbonnais told me that, in our future, it's possible that something as seemingly far-fetched as cloud seeding could be used to suppress lightning. But as we chatted, I learned that perhaps the most important question isn't whether we can suppress all lightning-related fire, but whether we should. As Bourbonnais put it to me: "Fundamentally, it does sort of precipitate this idea that fire is just bad, and that's not the case."

Wildfires are, after all, a natural and necessary part of many forest ecosystems, and our obsessive elimination of healthy fires has helped create the crisis we now face. Had fire been allowed to naturally play its role in Jasper's Athabasca Valley for the past hundred years, the monster that swept through in July 2024 would likely never have happened. Continuing to exclude wildfire from our landscapes only compounds the problem. Every lightning fire we avoid today just kicks the fire debt further down the road. Eventually Mother Nature will come to collect, often with more devastating consequences.

The problem with the current wildfire tech boom is that it's eating up funding and attention that could, and many argue should, be spent on the less flashy, more pedestrian prevention and mitigation tools we have known about for decades but seldom seem to use. Even Bourbonnais's networked microweather stations, as impressive as they are, were telling when it came to what all that data revealed about the McDougall Creek fire in the first place. No amount of wildland firefighting was likely to stop it.

Continually focusing on supposed silver bullet technologies feeds the unrealistic dream that we can engineer our way out of both the wildfire crisis and, more broadly, the climate crisis. But we can't code software to solve this problem, just like we can't suppress our way out of it. We should instead focus on relearning how to live with fire. The real answers aren't in some techno-utopian future, but in our past.

THE ONLY SOUNDS I CAN HEAR ARE CHIRPING BIRDS AND MY own footfalls as I weave along a narrow dirt trail through a phalanx of blackened trunks. The dead trees seem to march resignedly away up the hillsides to the horizon in every direction—tens of thousands of them, scorched and denuded of their needles and limbs. The sky overhead is steely grey, but as I round a bend in the trail, I'm greeted by an explosion of colour.

The forest floor is a carpet of green shot through with mock orange flowers and bursts of bright yellow heartleaf arnica. Dandelion fluff floats among the stalks of black on a gentle breeze, strangely reminiscent of the embers that drifted here two years before.

It was early May 2025, and I was on another reporting trip to the North Shuswap. Interviewing firestorm survivors can be emotionally heavy work, so I'd taken to stealing a few early morning hours to go trail running—a habit I picked up during the pandemic when it felt like trauma and fear were all around us. There are precious few designated hiking trails in the North Shuswap, an area where the mountainsides outside of town are a snakes-and-ladders checkerboard of logging cut blocks and forestry roads. One of the most popular trails starts from nearby Tsútswecw Provincial Park, where Ty Barrett and his crew huddled in their truck during the fire.

It follows the Adams River northwest towards Adams Lake. It's in this river valley that the Adams Lake and Bush Creek fires merged into one giant monster, rushing together and sweeping down the canyon. Just down the road is the airfield where four hundred firefighters made their chaotic escape through the flames that chased them to the Squilax gas station, the same gas station that exploded into a fireball as fleeing residents drove past.

I wanted to see up close the site of some of the most high-intensity crown fire of that devastating weekend, but what I found two years later was a tranquility so unexpected it left me standing dumbstruck in my trail shoes.

I've interviewed fire ecologists who have for years extolled the virtues of a fire-recovered landscape, more open and lusher and green with better forage for deer, berry patches for bears, roosts for birds. It sounded great in theory, but it had always been hard for me to square the visions of fire's destructive power with experts' descriptions of what tended to sound like Eden. Seeing it in person, in full spring bloom, is hard to describe without falling back on well-worn mountain writing cliches.

An eagle literally soared overhead, and sparrows squabbled around a nest in the crook of a giant, burned-out Douglas fir. Across the river, the hillside was carpeted in yellow, white and green. Even the blackened tree trunks seemed to glisten with a stark, arresting beauty. Somewhere in the distance, a train's whistle echoed through the mountains.

As I kept running, the river narrowed to a tight canyon rippled with rapids. At one particularly sharp corner, a set of stone-and-concrete steps led down to a rocky beach and the burned pieces of a fence and handrail. A scorched sign pointed out the remains of pictographs painted here by Secwépemc people thousands of years before European settlers arrived: evidence that people have thrived in this fire-prone landscape since time immemorial. The pictographs are hard to make out—not much more than red smudges against the grey rock—but they've survived wildfires before and will survive others yet to come.

This is the wildfire the public hasn't yet come to know. The rejuvenator, the bringer of life, the much-needed reset for forests that have grown too dense and thick and tangled.

ASHLEY O'NEIL KNOWS THIS KIND OF WILDFIRE. SHE GREW UP on her family's traditional territory north of present-day Cranbrook, BC, hearing stories of *good fire*, and how the Ktunaxa people stewarded their homelands with it for thousands of years. Wildfires are "a blessing in disguise," O'Neil said. "The land knew that we needed to restart."

I first met Ashley in April 2023. Clad in a blue Nomex jumpsuit made by her company, Ash Firewear, she was dragging a drip torch through dry grass along a logging road near the Cranbrook airport. Along with making wildfire gear specifically for women, O'Neil has founded several other wildfire-related companies, including a contract firefighting crew with twelve Indigenous firefighters. At Cranbrook, tendrils of fire trailed behind her and raced through the grasses in a widening triangle reaching deeper into the forest, building a crescendo of fire that engulfed shrubs and bushes, piles of fallen logs and branches. Here and there it reached into dry understory branches of trees and climbed them like a ladder up into the canopy, sending towering columns of fire shooting into the sky as individual trees candled and died. A Bell 206 helicopter clattered overhead, sprinkling incendiary ping-pong balls deeper into the forest, sparking little circles of fire that grew slowly out towards each other, adding to the growing blaze.

O'Neil was part of a 1,200-hectare prescribed fire project run jointly between the Ktunaxa community of ʔaq̓am, the City of Cranbrook, and the BC Wildfire Service. This burn day in April was the culmination of five years of work, one of the many prescribed fire projects that wildfire ecologist Bob Gray has helped design. O'Neil was working to earn her burn boss credentials, a requirement to oversee prescribed fires of her own one day. "When I was there watching it burn, it was relief for me, because I drive up there every day, I see it every day, and I see the benefits of that fire, what it had done in that area, the flowers, the medicine plants that we have gotten back," O'Neil said.

The goal of the project was to thin out the forest and to reduce the available surface fuels in a twelve-square-kilometre plot of land filling a triangle-shaped corner between the Cranbrook airport to the east, and Aq'am's Kootney 1 reserve to the south. Organizing a prescribed fire like this takes time, expertise and an incredible amount of human labour. First, the forest had to be analyzed and studied. Gray undertook that work in consultation with elders from the Aq'am community, who helped describe for him what that forest had looked like decades ago when it was healthy.

"We sort of picture BC as wall-to-wall conifers," Gray said, but that picture is flawed. We've come to view dense forests as healthy because the changes to them have happened so slowly, we struggle to perceive them ourselves. What we see today is what we think of as normal, but it's not. Somewhere between 40% and 50% of BC's forests, for example, shouldn't be wall-to-wall forests at all. They should be a mix of grasslands or shrublands, a mosaic, not a blanket of trees.

Imagine being able to climb onto a horse's back, pick any compass bearing you wished, and ride off between the trees in that direction with ease—that's how it was across much of BC only a few generations ago. Contrast that with the view in much of BC's forests today: woods so dense and dark and primordial you risk getting lost if you step even a few metres off a marked trail.

If we're going to have any impact on these large-scale, high-severity fires, we must start to see extremely dense forests as unhealthy, Gray said. The goal of prescribed fires and fuel treatments is not to prevent a wildfire from happening, Gray explained, it's to change the severity of the fire and alter the impacts of those fires: taking a landscape-sterilizing rank 6 fire with thirty-metre-high flames and turning it into a rank 2 or 3 ground fire—one that burns up dead litter and thins out the undergrowth, leaving larger mature trees behind.

The problem, Gray said, is the feedback loop we've created in heavily fuel-loaded forests. Fires burning through a fire-reliant forest every ten to sixty years encounter mostly fine fuels like grasses, needles and dead leaves. They consume those fuels quickly and burn fast, but with less intensity. Such fires can—and in fact must—happen on repeated cycles to keep the forest itself healthy. They're also not the types of wildfires that become stand-destroying monsters generating thousand-degree heat. But fires burning through forests that haven't seen wildfire in a hundred years encounter much heavier fuel load—more of the hundred- and thousand-hour fuels like branches, logs, stumps and whole trees. When a fire finally does burn all that additional fuel, it results in much higher fire severity. What would have been a healthy fire becomes an unhealthy one that nukes the landscape. In some cases, the forest may never fully recover. The best thing about prescribed fires, Gray said, is that they protect the forest from these devastating larger fires by allowing the cycle of healthy fire to return, even after the original treatment work has been done.

But writing a prescription is no easy feat. After the forest has been analyzed and studied, the next step is to figure out what natural fire's relationship to that landscape would be, and how to reintroduce it. Then you must consider weather and the overall goals of the prescription itself. Once that's all done you have to jump through dozens of bureaucratic hoops to get the government sign-off needed to proceed.

Colleen Ross is a wildfire ecologist who worked with Gray to design the Aq'am prescribed fires and served as burn boss on ignition day. She's also a former wildland firefighter, and she knows BC's forest landscapes intimately. There is, she says, no copy-and-paste formula. Every ecosystem is different. Though most of BC's forests have evolved with fire, some see natural fire return intervals as short as ten to twenty years in certain ecosystems, and sixty years in others. In the province's north, a forest might go two hundred years without seeing a naturally occurring fire. Across Canada, the situation is even more varied.

In Canadian forests, there are sixteen different wildfire fuel classification types, and an unlimited combination of fuel, weather and topography. Each permutation will respond to fire differently, with some landscapes more evolved to rely on fire than others. Most of Canada's Western forests, including the boreal and montane are *fire-adapted*—meaning they've evolved to rely on wildfire to help renew and rejuvenate them. But the role of wildfire in healthy landscapes is often far more complex, and therefore harder to recreate, than simply burning everything down and letting it grow back.

Understanding how fire interacts with all of Canada's varied forest landscapes is complicated, challenging work. Thankfully, we don't have to start from scratch. The land's original inhabitants have known for centuries what fires these lands need and how to provide them.

Indigenous communities learned over thousands of years how to use fire to cultivate forests and keep them healthy. But colonizers stamped those fires out. British Columbia was the first, in 1874, to outlaw cultural burning. By the mid twentieth century, the practice was illegal across the entire country (in many cases, along with other practices like potlatches, traditional dancing and even Indigenous languages). We taught ourselves to hate fire, to fear it, and firefighters became experts at suppressing virtually any fire that managed to start in a forest. We got so good at it that for over a hundred years we've effectively excluded fire from our landscapes altogether.

The result is that forests like ponderosa pine and Douglas fir—traditionally so open you could walk freely among the towering trunks—became much denser and more overgrown. Now when they burn, they often do so with an unnatural ferocity, raging so hot they can sterilize the soil and destroy every living thing, trees and all, leaving greasy hydrophobic soil behind.

Contrast that with forest types like the northern boreal forest, where lodgepole pine and black spruce naturally go eighty to two hundred years without seeing a fire. When those fires do burn naturally, they don't just scorch the ground. They burn in massive, high-severity blazes called stand-replacing fires, killing the majority of the trees, eliminating the canopies and leaving something closer to grassland in their wake. But this is also by design. These fires help preserve open spaces for deer to forage, for red-tailed hawks to patrol from above. And when the forest grows back, it does so quick and green.

The reality of the boreal forest today, however, is that—like its southern neighbours—we've managed to significantly reduce the number of fires it sees. That, combined with the impacts of climate change, has left it piled with fuel and ready to burn further, faster and hotter than Western society has ever seen. As Gray explained to me, these forests will burn no matter what we do—they need to. The best thing we can do is help that happen in the safest, least-destructive way possible. But the work that will take across the country is enormous.

Because our forests are so overgrown, you can't just wander into them and strike a match. Once you've assembled all the data to write the prescription, the forest has to be prepped. At Aq'am, the forest was so loaded with fuel that simply setting it alight would have created a fire that burned far hotter and more devastating than the prescription required. The fire could also easily have escaped and destroyed nearby homes or even the airport itself. That meant mechanically thinning the forest—selectively logging it, essentially—and cleaning up enough of the leftover slash to prevent the eventual prescribed fire from getting out of control. It also required buy-in from the community and a shared understanding of the story of wildfire on that landscape.

At the Aq'am burn, Ktunaxa Elder Marty Williams described how, though the prescription took five years to design and prepare, the community had been talking about returning fire to the landscape for decades. On the day of the burn, Williams welcomed dozens of firefighters, ecologists, pilots and other support staff with an opening prayer. He talked about how over his lifetime, the elk and deer his community historically relied on for food had been driven from the landscape by logging, agriculture and a forest that became so overgrown the animals could no longer navigate it. His hope was to see those animals return.

For Ross, these kinds of stories are vital, not just for ensuring that the prescriptions she writes are accurate for a landscape's needs, but for helping ensure everyone involved has a say and a stake in the process. "We want to be good stewards," Ross says. "I like to say it tells the story of what we're doing across time and space."

Finally, a successful prescription requires perfect timing. Like any fire, prescribed fires can be dangerous. To do them safely requires a Goldilocks combination of temperature, humidity, precipitation and wind—what prescribed fire practitioners call the *burn window*. Prescribed fires are most effective and safe during short periods in the spring when winter snows have receded, but before grasses, shrubs and trees have woken from their winter slumber and begun to green up; or in the fall after the summer's heat has dried out the fuels, but autumn temperatures and returning rains create a hedge against the fires exploding out of control. In both cases, Mother Nature must cooperate—too little rain and the risk of fires escaping becomes too great. Too much, and the fire won't burn enough to accomplish its goal. Sometimes you must wait months or years for the right burn window to arrive.

In April, the conditions aligned perfectly for Aq'am Elder Max Andrews to drip the ceremonial first flames onto the parched, prepared landscape, followed closely by a crew of BC Wildfire Service firefighters, the helicopter overhead, O'Neil and several students under Gray's tutelage. As fingers of flame crawled into the forest that April morning, nobody knew that only a few months later, their burn would face its ultimate test.

WHEN THE ST. MARY'S RIVER FIRE ROARED TO LIFE JUST AFTER 1 p.m. on July 17, 2023, O'Neil knew immediately that it posed a dire threat. Having worked on the Aq'am prescribed burn that spring, she knew these forests well. She knew how heavy the fuel loads were, and how dry.

She arrived at the fire within an hour of its start and met Cranbrook Fire Chief Scott Driver, who'd also just arrived. Both recognized that dozens of homes were at risk. The fire was ripping towards the community, feeding on bunchgrass dried and cured by the summer's high temperatures and crippling drought. It climbed the ladder branches of towering ponderosa pines, torching through their canopies and showering embers ahead of itself on sixty-kilometre-per-hour winds. O'Neil and other local residents teamed up with Driver's firefighters from Cranbrook and raced door to door, telling residents to get out. The fire destroyed seven homes in Aq'am within hours of starting.

Days later, when the threat to homes had subsided, O'Neil drove into the fire zone above the reserve to scout where the fire was burning. She drove up the steep cliffside road towards the triangle-shaped patch of forest she'd dragged a drip torch through just a few months earlier. She watched as the fire's embers from the untreated side of the road arced over her head, landing in the area they'd burned in April.

"It got into the prescribed fire part, and it just slowed right down," O'Neil said. "They stopped it right at the airport." The prescribed burn had done its job. Without that buffer, O'Neil figures the reserve would likely have lost three hundred homes, instead of just seven. The Canadian Rockies International Airport in Cranbrook isn't just a local airstrip; it's a critical hub for moving wildland fire resources around the province to respond to other fires. Losing it could have crippled wildfire response across BC's entire southeast.

Preventative burns can also speed up the time it takes for a forest to recover. Even months after the St. Mary's fire, the impacts were obvious. Inside the prescribed fire's boundaries, trees were scorched but alive, with new shoots of grass and shrubs already pushing up through the singed forest floor. In the untreated areas, the forest was blackened beyond recognition, with most of the trees dead.

AMY CARDINAL CHRISTIANSON DIDN'T GROW UP AROUND CULtural fire. Like many Indigenous people in Canada, her family's relationship with fire and the landscape was severed by colonialism. But while she was growing up in a Cree-speaking Métis family in Treaty 8 territory in Northern Alberta, wildfires were never far away. In Northern Alberta, fire is and always has been a routine part of life, even after cultural burning practices were largely stamped out. And while she didn't grow up with "a match in [her] hand," Christianson did start her journey with fire from a more comfortable place than most of us.

Today, Christianson is the senior fire advisor for the Indigenous Leadership Initiative. Before that she was a research scientist for the Canadian Forest Service, and an Indigenous fire specialist with the National Fire Management division of Parks Canada. This all gives her a unique ability to speak multiple languages across often-divided fire lines, from Western science to Indigenous culture and tradition. Though she's been a leader in the space for over a decade, Christianson's relationship with fire is also still evolving, rekindling over time. "I'm on a journey of reconnection with fire and have been really lucky to have a bunch of Elders who've taught me and supported me," she said.

Christianson's first introduction to cultural fire began 20 years ago during her PhD work on fire hazard management with the Peavine Métis settlement, where she learned about using fire as a way of cleaning the land. Her work had originally focused on more Western concepts like fire-smarting and fuel treatment around houses, but Elders pointed her to a larger truth. "They said, 'Our landscapes are unhealthy, and unless we clean the land, we're going to continue to experience these fire problems.'"

To understand more about how we got to this unhealthy landscape, it's helpful to go back to the dawn of Canada's forestry industry. The holdovers of colonial thinking from this era are still with us today.

The beginnings of professional forestry in Canada mirror those in the United States, driven largely by settlers' insatiable demand for timber. During the late 1700s and early 1800s, Britain relied on Canadian forests for everything from lumber for houses to towering white pine trunks that became the masts of its ocean-conquering navy. During the height of the Napoleonic wars, the British navy required between four thousand and six thousand oak trees to build just one of its infamous hundred-gun man-of-war ships. As timber supplies were strained across the British Isles, the Admiralty looked to North America, but it was hardly the only colonial power hungry for trees. Whole cities in the so-called New World were being built largely out of timber. But as first hardwood and then softwood forests were razed across Eastern Canada, alarm began to grow.

By the 1870s, Canada's first prime minister, Sir John A. Macdonald, was worried. "We are recklessly destroying the timber of Canada and there is scarcely a possibility of replacing it," he wrote in a letter to Ontario's premier at the time. The new country needed a way to protect and preserve its forests.

In 1899, Elihu Stewart was appointed Canada's chief inspector of timber and forestry, and the roots of today's forestry industry really took hold. Stewart envisioned a future where forests were protected and stewarded by professional foresters—a designation that did not yet exist. "The whole forestry problem in our Northwest may be concluded in two words: conservation and propagation," he wrote. "Conservation," in Stewart's mind, "involves preventing destruction by fire, and secondly, a judicious system of cutting the timber so as to retain for all time a continuous supply."

In Canada, Stewart oversaw the founding of a fire guarding program that hired local settlers as fire rangers who provided their horses for $3 per day. By 1904, they'd all but perfected the art of suppressing wildfires. A similar explosion of firefighting was unfolding in the US. After a devastating series of fires in 1910, known as the Big Blowup, the US Forest Service implemented what became known as the 10 a.m. policy. It mandated that every wildfire be extinguished by 10 a.m. the day following its discovery, and was quickly adopted across Canada as well.

This utter obsession with putting out all fires accomplished only one thing. It created what fire ecologists today call the wildfire paradox: The more money and effort that is poured into fighting wildfires, the deeper the wildfire debt on the landscape and the worse the next wildfires will be. The harder we battle fire, the worse the problem gets. Seen that way, the devastating fire that destroyed a third of Jasper in July 2024 wasn't an aberration; it was the bill coming due for a century of healthy fires that we refused to let burn.

Christianson told me Indigenous communities have known all along that we can't firefight our way out of this problem. "Unhealthy fires will happen on unhealthy landscapes," she said. Until we correct the underlying disease, the symptoms will keep getting worse.

While experts like Christianson are working to change how we think and talk about wildfires, she can't help but see the hand of colonial thinking reaching from Stewart's era through to today. One of the clearest examples in her mind is Canada's increasing centralization of wildfire management. In British Columbia, the practice of organizing firefighters into twenty-person unit crews began decades ago as a way to engage rural, mostly Indigenous community members because they were the workforce closest to the fires, and most available to help manage them. But over time those rurally oriented community crews were replaced with specialized professional firefighters organized by centralized command structures and dispatched to fires hundreds or thousands of kilometres away from where they live. It made sense from an efficiency and safety perspective, but it also divorced wildfire management from the people who live where the fires are happening.

"Traditionally, for Indigenous people, you managed and stewarded the land and had a relationship with fire where you lived," Christianson said. "It was like a long-term relationship. And so, what we've seen now through this centralization of fire management is that lots of people who are making decisions about what happens on a fire sometimes haven't even visited the areas before, which is always shocking to me, yet they're overruling local people on the land."

It's no surprise to Christianson that conflict arises in communities that have been culturally separated from fire and then dictated to by centralized government agencies who don't know the land like they do.

And yet a frequent refrain from politicians searching for an answer to the crisis isn't to put knowledge, agency and resources back in the hands of local communities. It's to demand a national wildland firefighting service, akin to the military, that could parachute in from across the country and battle wildfires with presumably unlimited resources. Surely the answer, many argue, is more manpower, more water bombers, more insistence that we must fight these fire-breathing beasts to the death.

After the 2021 wildfire season, the Canadian government promised to fund one thousand more firefighter positions nationally. After the 2025 season that took a heavy toll on Manitoba and Saskatchewan, calls grew louder. Manitoba premier Wab Kinew, Newfoundland premier John Hogan and many others called for Canada to invest billions to build a new fleet of water bombers. Even A-list experts like Thompson Rivers University's Mike Flannigan have called for a national wildland firefighting force. And while that idea might help during especially bad fire seasons, the most destructive wildfires now routinely burn so fast and hot that no amount of firefighting resources—federal or otherwise—will stop them. What we need to do instead is adjust our relationships with fire, and that includes relearning how to live with it. It's here that Christianson sees Indigenous knowledge leading by example.

She has been working with the Kainai Nation, also known as the Blood Tribe, near Alberta's Waterton Lakes National Park, to help build a truly organic fire guardian program, the first of its kind in Canada.

The goal, Christianson said, is to help bring fire back to the landscape in sustainable, healthy ways. Fire shouldn't be seen as something bad that happens in the summer, but something necessary that takes place almost all year long, a vital part of communities themselves. In many ways, it's not even really about fire. "It's about healthy cultures, healthy land and then a healthy people" in a landscape that includes fire.

But achieving this takes time and, most importantly, money.

Spending on fighting wildfires directly now regularly exceeds $1 billion per year, but less than 10% of that gets spent annually on prevention. Between 2002 and 2023, British Columbia, with the country's largest standing firefighting service, spent $6 billion fighting wildfires, and only $300 million on prevention and mitigation. There is somehow always money for water bombers or out-of-country firefighters, or even a potential federal firefighting force, but only relative pennies for prevention or Indigenous partners.

As frustrating as these current realities are, they are slowly shifting, pressed forward not by political will but by reality. The type of fire seasons we now frequently face can't be managed within our existing systems. We need a new approach, one that leverages all of our communities' combined strengths.

CHAPTER 12

COMMUNITIES AT THE READY

A cold rain splatters my windshield as I scan the gravel roadside for the driveway Terry Jessup told me about. Ahead of me, an inky storm cloud rises over the rolling cattle pastures of Knutsford, BC, in almost the exact same place a giant smoke column blotted out the sun as I drove these same roads nearly two years before. Back then, in August 2023, I was covering the Rossmoore Lake wildfire, a 114-square-kilometre fire that for almost a month threatened the ranching community of Knutsford and the roughly 100,000 residents of Kamloops barely twenty kilometres north of here. By early August, the fire was so big it took almost an hour to drive its perimeter along well-worn logging and resource roads cut through the hills and forests.

Now, in May 2025, I'm searching for Jessup and his crew from the Knutsford Initial Response Team, or KIRT, a group of community volunteers from this rural area that banded together to fight wildfires themselves. I'd spent the past two years interviewing dozens of firefighters and talking with fire behaviour experts, Indigenous fire stewards and local residents who, like Jessup and his crew, wanted nothing more than to be part of the defence of their communities. I'd learned almost everything I could about some of the most intense and destructive Canadian wildfires of the past five years. Before all of that, I had travelled to rural Australia, trying to understand if there were ways we could fight wildfires differently. The sum of all that research had led me here, to an abandoned-looking ranch house nestled in a saddle between Knutsford's rolling hills.

The front door is ajar and guarded by two large horses who'd apparently wandered over from nearby pastures. I approach the horses gingerly, clicking my cheeks so they don't startle at my presence. I really don't want to get kicked.

"C'mon in!" Jessup's shout echoes from somewhere inside. I politely excuse myself to the horses and step inside, where I find Jessup wrestling with a cranky ink-jet printer in a back bedroom of his de facto headquarters, trying to get radio and incident command manuals printed before all his firefighters show up for their twice-monthly training nights.

The house, which a community supporter allows them to use for free, has a threadbare carpet and no running water. A giant Australian flag is pinned to the living room wall, with South African and Costa Rican flags hanging nearby. The flags were gifts from firefighters Jessup and his crew wound up working alongside during the Rossmoore Lake fire. There were also gifted boxes of used protective fire equipment like helmets, goggles, Nomex shirts and pants, all the stuff that Australian firefighters had brought with them and were unable to take home. It's great to have on hand, Jessup says, and a welcome show of camaraderie from fellow "firies" on the far side of the world. He's grateful for the gesture, and the bonds his crew built with foreign firefighters during the worst wildfire season in Canadian history. To him, the shirts and helmets hanging in this small bedroom closet represent KIRT's acceptance into the global wildland firefighting family.

Before long, other members start to arrive, each making polite if careful greetings to the horses at the door as they enter. There's Craig Palmer, Jessup's captain and second-in-command. Karl Thorson, in a blue cotton hoodie worn under his red KIRT-branded Nomex coveralls, greets me with a nearly crushing handshake. Next comes Mike Connolly, another captain who owns a ranch near Rossmoore Lake, and whose property was at risk almost the entire time through the fire. Lea Thorson, Karl's mother, walks in with a bright smile. Before long, the empty living room is filled with fire-resistant suits. Next to their safety boots is a stack of hand-me-down piss-can backpacks with K.I.R.T. inked across the top in permanent marker.

Jessup passes around the newly printed manuals and together they dive into the details of radio communication policy and etiquette, of incident structures and chains of command. There's even a page detailing each of the terms in the phonetic alphabet; the *Alpha, Bravo, Charlie, Delta* and so on that soldiers—and firefighters—use when communicating under fire over patchy, static-choked radios. They go over how to make an incident report to BC Wildfire using this army jargon. Perfection, it seems, takes practice.

"Don't worry if you can't remember all of it right away," Jessup says. The point right now is to know these procedures exist, and where to find them in the manual when you need them.

After about an hour of reviewing policy and procedures, the crew heads back outside for some hands-on drills. The setting sun has managed to peek under the storm clouds, refracting through fat raindrops to cast a rainbow against the nearby mountains and bathing everything in a warm, golden glow. Jessup and Palmer start sketching out the parameters of their exercise.

A new "wildfire" start has been reported in the grass and shrubs behind the house-turned-headquarters. The crew has one thousand-litre tank and a pump, thanks to the trailer that Bernie Fink brought with him for the training. The job will be to run out a section of hose, split it in two and surround the imaginary fire, attacking it from the flanks while being careful not to tread in front of the head. At first things go smoothly. Karl Thorson and Aron Wiens—a burly man with a close-cropped mohawk and closer-cropped beard—start stringing out hose line while Fink walks Lea Thorson and a few others through his home-designed pump set-up. As the pump itself clatters to life, the sound is deafening, and everyone clamps their hearing protection over their ears.

In the field, Wiens is struggling to unfurl a melon of tightly coiled firehose. With some straining and cursing under his breath, he manages to free the bundle and connects a three-way valve, splitting the hose line into a Y shape. Thorson takes one end and runs it up a small ridgeline above the fire, which they're told is burning in a stand of aspen and berry bushes behind the house, trying to ignore the rainwater soaking through his boots from the grass. Wiens then takes the other end and circles around below the fire. Once set, they both call over their radios for water from the pump. At first, nothing. Then, yes, something. A spurt of white from Thorson's nozzle, then a hiccup, followed by a steady stream arcing out into the bushes. Wiens is again struggling with his line, trying to figure out how to handle the awkward ninety-degree valve that Fink has left attached at the end. It's an irrigation hose, not a standard firefighting one, but soon he has it figured out, sweeping a jet of water in wide, flowing arcs.

Back at the water pump a hundred metres away, Jessup calls into his radio for each crew member to check in. Some do, but a few don't, so he tries again. A better response this time, but still not everyone, so he and Palmer head out into the field to touch base in person. There's no real risk right now—Jessup can see everyone from where he's standing—but the point is to test their abilities with the radios and their communication systems. In a real fire, with smoke and embers swirling, it can be easy to lose track of crew members, and failing to respond to radio calls can easily lead to a "firefighter down" mayday call. As the scenario winds down, everyone regroups at Fink's pump trailer for a debrief.

"It could have been better," Wiens says with a rueful shrug. "I'm not super confident yet, I just need more practice." The irrigation hose fittings were a particular surprise, he says. He was expecting standard firefighting hardware.

"True, but that might happen in the real world," Jessup says. "You just have to figure it out." Their work with the radios also wasn't perfect, but it was better than the last time they ran a simulation like this, and that's the point: constant improvement.

The day after the KIRT training, I met up with Thorson again near his family's ranch, and we drove around touring the burn scar of the Rossmoore Lake fire. He showed me where he and his crew had held the line on multiple days, where they'd done important mop-up work in some areas to allow professional firefighters to focus on more challenging tasks elsewhere.

As Thorson's pickup weaves along the logging road, we reach a junction that I recognize. I'd stood here two years before as everything west of the road burned and firefighters fought to keep the fire from crossing. Mexican firefighters who'd been flown in to help bolster BC's struggling resources had worked their way into the smouldering forest from the roadway, swinging dual-wielding machetes as they went, chopping the lower branches from the spruce and pine trees standing among the towering Douglas firs. They were so efficient several BC firefighters asked if the province should start equipping its crews with machetes as well. Then the Mexican crews burned the area with drip torches, using fire to eat up the surface fuels and help secure the line they'd built along the logging road.

Seeing it two years ago, amid the smoke and flames and 30+ degrees Celsius heat, it looked like you'd expect: a disaster zone. But driving through it with Thorson, we were both struck by how it looks today: like a healthy forest. Flowering greenery carpeted the forest floor, and the tree canopy was pushed three metres off the ground. Even most of the spruce and pines that the Mexican firefighters had attacked with their blades had survived, and those that hadn't now left holes in the overhead canopy where sunlight streamed in.

From where I stood on the road, I could see that the firefighters' efforts had worked, and that the forest behind us to the east had not burned. It was still a dense, dark tangled mass of brambles, fallen trees, dead shrubs and so much fuel. Looking at it now, in contrast to the old fire beside it, I watched Thorson's understanding of the forest where he lived begin to shift in almost real time, from seeing it as I once did, utterly ordinary, to how I see it now: an overgrown mass of fuel just waiting for a spark. A lot had changed for him since this forest first caught fire.

IN LATE JULY 2023, THORSON SAW THE LIGHTNING BOLTS THAT sparked the Rossmoore Lake fire. They shattered the sky, crashing down into the dense ponderosa, lodgepole pine and Douglas fir forests in the hills around his family's ranch. They hit in multiple places, striking trees and high ridges on the horizon around him as the storm rolled through.

In the rangelands just south of Kamloops, BC, rolling open pastures meet the towering trees of the province's higher mountains. In fine weather, you could imagine Frederic Remington, paintbrush in hand, conjuring hard-bitten cowboys corralling cattle under an endless sky. But in July 2023, the surrounding forests were tinder dry, and Thorson knew it. Nervously he scanned the horizon for any sign of fire and at first saw nothing. He started to relax. The lightning storm brought rain with it, and Thorson assumed it would quell whatever starts the lightning might have ignited.

Then his neighbour called to say he'd spotted smoke in the hills to their south. "Oh shit," Thorson muttered aloud. He turned to his mother, Lea, both of them immediately worrying about their animals. Dreamscape Ranch isn't your typical cattle ranch. It's essentially a retirement home for horses: forty-five of them that, after long lives as working animals, now had a home with open pastures, attentive care, and a calm, natural environment. It's also home to one hundred chickens, four cats and three aging dogs, not to mention the half-dozen staff who help run the place. All of them would have to be evacuated if a fire threatened the ranch. It wouldn't be easy.

Lea started calling friends and neighbours, anyone with a livestock trailer who might be able to help. Karl hopped on the ranch's side-by-side and drove up into the forests, trying to scout out the fire's exact location. Weaving along cattle trails through the hills, it took Thorson what seemed like forever to find it. As he searched, he drove deeper into a forest so loaded with fuel it would be nearly impossible to walk through. Tangles of dead trees and branches littered the floor like piles of skeletons, sometimes chest-deep, their bones bleached grey by years of drought. Dried out brambles ran like coils of barbed wire through the forest, snagged here and there in the dead, brittle lower branches of the trees themselves. From the forest floor to the towering canopies above, it was an almost solid wall of fuel ready to burn.

At first, when Karl finally located it, the fire seemed far away. "Oh, it's way over by Lac Le Jeune. We'll be okay," he thought as he pointed the side-by-side towards home. But not long after, the fire took off, racing through the dense forest Karl had just driven. It sent a towering column of smoke and ash into the sky, casting a shadow over downtown Kamloops, more than twenty kilometres away.

Hours or possibly a day later (Lea's memory of that frantic time is a blur), RCMP officers came to the ranch and told them to prepare to leave. The ranch's quiet and calm dissolved into scenes of anxiety and chaos almost overnight. Horses who'd spent years lounging in fields studded with wildflowers now had to be crammed into claustrophobic livestock trailers they hadn't seen in a decade and taken—somewhere.

"Where do you take forty-five horses?" Lea wondered at first, growing stressed. But if there's one thing rural communities in Canada are good at, it's coming together in a crisis. Before she knew it, trucks and people were arriving from near and far. "We had trailer after trailer," Lea said. They worked into the night. Cowboys she'd never met showed up to help, lending their skill and experience to aid in loading the skittish horses into trailers as the sun set behind the ever-growing smoke column. Leaning on friends and neighbours, Lea was able to get all her horses safely to another ranch four hours away in Valemount, BC.

As Lea raced to coordinate the animals' evacuation, Karl and other locals started trying to fight the fire directly. For the first few days, it was madness, he said. Most had no idea what they were doing, and they made the kind of simple but potentially fatal mistakes that people who've never fought a major wildfire before are likely to make: like driving a convoy of pickup trucks down a cattle trail into the fire without stopping to consider escape routes. Or then parking those trucks facing into the fire instead of out of it. "We did some stuff that was really, really dumb," Karl said, looking back. "I was the guy in sandals at one point, thinking I'm invincible."

This is exactly the kind of scene that infuriates professional wildland firefighters: well-meaning but stubborn locals racing hither and yon in their redneck firetrucks, spraying water on anything that looks like a flame with no training, no tactics and little understanding of how to meaningfully fight a wildfire. At first, this is exactly what Karl and his neighbours were doing, and while they had moderate results in keeping the fire at bay, their activities were also extremely dangerous.

"Everybody wanted to help, but we weren't really all that coordinated or sure what to do," said Gord Peterson, one of Karl's neighbours. The instinct to help is understandable and human, but preparing to face an oncoming wildfire isn't something you can do at the last minute. It takes training, organization, equipment and no small amount of paperwork: the kind of boring, glamourless work that a lot of people (busy ranchers among them) often can't be bothered to do in the quiet months between fire seasons. When flames are threatening your home or livelihood, the pressure to not only do something but *feel* like you're doing something is profound. But just as government agencies and policy-makers often suffer from "rainesia" after the fire season has ended, so too do many of the community members who demand access to fight the fire at the height of a crisis. After all, setting up societies, arranging training and insurance, and filling out Canada Revenue Agency paperwork is precisely the kind of mind-numbing bureaucratic hoop-jumping that many rural Canadians hoped to escape by living away from cities in the first place. Karl, Gord and their neighbours weren't yet part of KIRT. They were, like many rural folks when a wildfire strikes, just trying to do their best. Though they didn't know it at the time, there were in fact two people who'd already started doing all the tedious legwork necessary for volunteers to fight fires properly.

YEARS BEFORE THE ROSSMOORE LAKE FIRE, PALMER AND JESsup realized the need to get organized. They'd seen half-hearted attempts to get rogue community fire brigades going before, and it usually didn't end well. Jessup in particular remembers hearing about the roadblocks, both physical and bureaucratic, that other well-meaning folks had run into on past fires.

Palmer spent much of his life as a structural firefighter in Kelowna, BC, while Jessup worked forty years as an advanced care paramedic. Both of them worked for decades alongside other first responders. They understood incident command structures, were cool under pressure and—most importantly—they could speak the same emergency response language as the BC Wildfire Service.

Two years before Rossmoore Lake, Jessup had helped create the fledgling volunteer firefighting group that some jokingly referred to as the Knutsford Militia. In 2021, an arsonist had been going around setting grass fires in the area. With no municipal fire service to speak of, and provincial firefighting resources stretched under what was then the third-worst fire season in BC history, the ranchers and farmers of Knutsford, BC, had little choice but to fight the fires themselves.

When lightning struck the nearby Juniper Ridge that summer, sparking a fire and threatening both a Kamloops subdivision and many Knutsford properties, the community held a town hall meeting with representatives from the local Thompson-Nicola Regional District. Jessup used the meeting to pitch the idea of a locally run volunteer fire brigade. He was greeted by a resounding "No."

"The regional representative at the time told us that they were not interested in funding a volunteer fire department. They wanted nothing to do with it," Jessup said. "That's when a few of us decided that, well, we've got to do something."

Without the support of their regional government, the ragtag group started organizing anyway. They hired an instructor to deliver the basic wildland firefighting course called S-100, and basic fire entrapment avoidance, called the S-185. Around thirty people signed up. In 2021, hardly anyone outside of the BC Wildfire Service knew what S-100 or S-185 even meant. Today they're common refrains in fire-prone communities across the province, as locals start adopting the jargon and practices of the professional wildland firefighting world.

Community members banded together to pay for the training themselves. The two courses take about two days to deliver, mostly classroom time with some hands-on work setting up pumps and hose lines. It's not enough to make someone a fully-fledged professional firefighter, but it's a start. It usually costs around $500 per person. For a ranching community on a shoestring budget, $15,000 to cover training for thirty people isn't always easy to come by, but they managed to scrape enough together.

They also started organizing their own "redneck firetrucks"—half-ton pickups with thousand-litre water tanks in the back. The trucks and equipment can easily cost between $60,000 and $100,000, not to mention the water tanks, hoses and a pump. To keep costs down, most of the Knutsford folks outfitted their private vehicles for firefighting. Jessup's own truck is his "daily driver." Like many in his crew, he can often be found fetching the mail or doing grocery runs with a vehicle weighing more than 4,500 kilograms. A few crew members added their own chainsaws, hard hats, pickaxes and piss-can backpacks. Some folks had their own PPE; some didn't. It wasn't much, but it felt like the start of something bigger.

But just as things were gathering steam, rainesia struck again. The summer of 2022 was wet, with few fires to fight. Of the thirty or so people who'd completed the training, most drifted away. By the time the Rossmoore Lake fire kicked off in 2023, only three or four members of Jessup's crew were still participating regularly. "No fault on anybody, you know?" Terry said, looking back. "They had other priorities to deal with, their livestock, or whatever—lives got in the way."

Even so, it meant that by the time the 2023 fire season rolled around, Jessup's crew had dwindled, and he worried it might fizzle altogether.

THE LIGHTNING STRIKES THAT KARL THORSON WATCHED CRASHing down around his ranch in July 2023 didn't just spark the Rossmoore Lake fire. They also ignited the Scuitto Creek fire, about fifteen kilometres to the east. As Thorson was scouting through the forest around Rossmoore Lake, Jessup was driving towards the Scuitto Creek with a BC Wildfire Service initial attack crew hot on his heels. They arrived at the fire almost simultaneously. Jessup jumped out of his truck and quickly introduced himself. "Whatever's in my truck is yours to use," he told the crew of firefighters. "What do you need?"

The crew leader's answer was simple. They needed water.

"Alright," Jessup said, dropping the tailgate on his truck. Within minutes, the IA crew members were pumping water out of Jessup's tank to attack the fire themselves while he lugged his own pump down to a nearby pond and set it up to resupply the crew's water.

"I think where we excelled right from the get-go is that the rules of engagement were made clear to us by Wildfire, and we followed those rules," Jessup said, looking back. "We wanted to be there, we wanted to help out, and we knew that ability would be short-lived if we didn't comply." One of the most important rules that Jessup continues to live by today is that as soon as BC Wildfire Service firefighters arrive at a fire, it becomes their scene to control, and what they say goes. Jessup's crews are happy to oblige.

When he realized they'd need more water than his truck and

their combined pumps could provide, Jessup pulled out his cellphone and called one of his group's community members who knew the area better than he did. "Where's the nearest water?" he asked. The next thing he knew, another community member arrived driving a fifteen-thousand-litre tanker and parked it in a nearby field. Minutes after that, another one arrived. By the time a structural fire crew from Kamloops arrived on scene to help, the fire was well on its way to being contained.

At 6 that evening, mere hours after the fire started, the BC Wildfire Service declared the Scuitto Creek fire held at fourteen hectares. It was an example of near-seamless cooperation between professional firefighters and amateur ones, and though no official recognition was given, to Jessup it still felt like a win.

This kind of cooperation between professional wildland firefighters and locals is often the exception rather than the norm in Western Canada, particularly in rural communities that prize their independence and can often be distrusting of centralized governments. When it works, it relies on mutual respect and mutual trust. As fire seasons have worsened, and the impacts on communities continue to mount, examples of that trust breaking down the way it did in the North Shuswap, just sixty kilometres from where Jessup and his volunteers fought side by side with professionals, are increasing in number. Fire seasons will continue getting longer, and fires themselves more extreme. Trying to force communities to "obey" hasn't been working, which makes cooperation with local volunteers ever more critical.

More important than his truck or his pumps or the dozens of locals Jessup is able to mobilize, the ace up his sleeve is his ability to speak the same language as both firefighters and ranchers, to be the bridge between twenty-four-year-old fire crew leaders and sixty-four-year-old ranchers who distrust the government. When he first encountered that crew at Scuitto Creek, and in calls with the wildfire service afterwards, speaking a common first responder language allowed Jessup to lay the foundation for a relationship with the fire service that would prove critical in the weeks to come. As firefighters mopped up around the edges of Scuitto Creek fire, Jessup looked across the valley and saw the smoke column from the Rossmoore Lake fire rising into the sky.

GORD PETERSON WAS ON HOLIDAY IN ALBERTA WHEN HE FIRST heard about the Rossmoore Lake wildfire and the evacuation orders that were starting to roll out. When he saw the news, he didn't know if he'd be able to get back home in time or not, but he figured he had to try. His family packed up and drove through the night to get back to Knutsford. By the time he reached his property, which neighbours Karl Thorson's on Long Lake Road, the RCMP already had roadblocks in place. But Peterson explained who he was and that he desperately needed to reach his property, and the police decided to let him through with a warning: If he left the area, he might not be able to get back in again. "Right from the start, I think it was pretty important that there was no adversarial stuff going on. The people who decided to stay and fight the fire were respected," Peterson said.

Respected, maybe, but at first they were not particularly well organized. The early days of the firefight, when Thorson was wearing sandals, were particularly haphazard. He remembers the day Palmer and Jessup arrived on the scene as locals were fighting to defend a neighbour's property. Confusion and chaos reigned. Locals with redneck firetrucks had hoses running all over the place, disappearing into the smoke, with no apparent cohesion. There was no clear command structure, no organized escape routes or safety zones.

With so little organization, Jessup and Palmer urged them to leave. It had become too dangerous to stay.

After helping knock down the Scuitto Creek fire, Jessup and Palmer approached the BC Wildfire Service about getting their volunteers designated as statutory hires, bringing them under the fire service's official umbrella, providing pay and even insurance coverage. As luck would have it, the Rossmoore Lake fire was under aegis of Incident Management Team 5—Healey's team—and they were juggling the Adams Lake and Bush Creek fires as well. Based on what their crews had seen of Jessup's professionalism at Scuitto Creek, the wildfire service agreed.

At first, they were assigned mostly to mop up and other lower-risk jobs, freeing professional firefighters to focus on more stubborn parts of the fire. But Jessup says as the days and weeks rolled by, the fire service came to rely on them more and more, eventually assigning them to tasks alongside crews from Mexico, Australia and Costa Rica. The more they worked together, the stronger the trust between professional and civilian firefighters became, eventually resulting in the KIRT crews being trusted to do the kind of work any professional Type 2 crew would get assigned.

By mid-August, it looked as though the Rossmoore Lake fire was well on its way to being held. Thorson, Peterson and their neighbours began to believe they'd escaped the worst of it. Under Jessup's guidance, they had helped keep the fire mostly to the west of a large pipeline, and out of Peterson's heavily treed lots. But one evening the winds shifted and started pushing the fire east as darkness fell, right towards those trees and his house beyond.

Again, the Knutsford crews went to work, battling the blaze along the ranching trails and logging roads in the hills above Peterson's property. On one pivotal weekend, they fought late into the night. It was August 18—the same day that a cold front blowing down across the province supercharged the Adams Lake and West

Kelowna fires. Peterson said the BC Wildfire crews had laid out everything they'd need to try and hold the fire—hoses, pumps, fire guards, the works. But when those other fires exploded, threatening far more homes than Knutsford, the firefighters were suddenly redeployed. The situation was so dire across the region that Knutsford was, for the moment, essentially on its own.

When the winds hit, the fire took off, and another evacuation order was issued. This time when the RCMP officers arrived, they didn't try to enforce the order, they just delivered the news and asked Peterson what he wanted to do. The respect the Knutsford crew had earned with the fire service over weeks of work on the fire paid off. Rather than confine people to their properties or force them to leave, the police let them stay and attack the fire themselves. "Our guys didn't go home," Peterson said. "They were up there at 11 at night, and these fires were coming." Thorson and several others worked shrouded in smoke, silhouetted against pockets of flame on the forest floor, digging fire guard and running hoses. They managed to hold it at Peterson's fence line and likely saved his property. "It probably would have come right across that night," he said. "You can still see the scorch marks on the trees."

The Knutsford crew's efforts alongside professional firefighters at Rossmoore Lake were not free of disagreement. A significant part of their battle unfolded next to several giant slash piles in the forests near Peterson's property. Logging debris had been left stacked as high as a house in three places. If the fire got into them, it would be impossible to put out. Worse still, the years-old slash piles were like bushels of flaming arrows just waiting to be set alight and sent aloft. If they went up, they would shower embers right into the heavy timber on Peterson's property. That, Thorson feared, would trigger a chain reaction that could send the fire racing right over the ridgeline and down towards his family's ranch and other homes and properties beyond.

The Knutsford firefighters spent days defending those slash

piles, digging fire guard around them, and fighting off the fire with pumps and hoses, backed up with frequent helicopter water drops from the lake on Thorson's property. For days, the crew stubbornly refused to let the slash piles burn.

Thorson described several times debating with leaders from the BC Wildfire Service whether the efforts to protect the slash piles was worth the resources they were tying up. Perhaps they should instead try to find an optimal time to burn them on purpose with winds in their favour. But the locals always insisted burning the piles posed too great a risk to their homes, and—though not without some butting of heads—the fire service agreed to let them keep up the defence. Wildfires are by definition stressful times. No one expects every conversation to be the most pleasant they've ever had, but the wildfire service's willingness to listen, to hear Jessup's crew out and ultimately to trust them likely cemented an alliance for years to come.

Even two years later, the evidence of the fight around the slash piles can still be seen in the forest above Peterson's land. The piles are now just scattered embers surrounded by the deep, wide fire guards that the Knutsford crews helped build and hold. The piles were eventually burned under safe conditions long after the wildfire itself was contained.

There were other challenges too. Not everyone in the Knutsford community appreciated Jessup's willingness to play ball with the wildfire service. Some disliked being told what to do on their own properties and continued trying to fight the fires without cooperating. This isn't entirely surprising. The same rural self-reliance that makes many people so handy in a crisis can easily become the wedge that splits them apart. Eastern Canadians often like to think of BC as the home of hemp-wearing granolas, and on the coast I'll admit this isn't too far off. But rural BC bleeds deep conservative blue, and those politics don't always mesh well with a fire service that's headquartered in Victoria.

Thorson described one particularly frustrating moment where he and other KIRT firefighters were helping contain spot fires that had ignited across their lines in the hills above Peterson's property. Jessup was on the radio coordinating with a helicopter overhead doing water drops, ensuring his people were out of the way. Another local who'd rebuffed efforts to come inside the fold was also on scene doing whatever he thought was necessary. When they tried to warn the man about an incoming helicopter drop, the man refused to listen or get out of the way until the last second, risking getting crushed by a wall of water or forcing the pilot to abort altogether. It was the kind of close call that grates on Jessup's nerves as a career first responder. "I can honestly tell you, after forty years as a paramedic, if you were to get in my way, it's not going to go well for you," he said.

Despite the challenges, the cooperative agreements held thanks in large part to Jessup and Palmer's work not just as leaders, but as bridges between the local community and the fire service. As the fight ground on, the fire eventually came under control thanks to the monumental effort of all the crews—professional and amateur, local and foreign. But the work to build on what Jessup and the others had started didn't end when the Rossmoore Lake fire was declared out. In fact, it was only just beginning. With 2023's successes notched on their belts, it was time to truly formalize their organization.

To keep doing what they were doing, Jessup and Palmer needed more formal structures. They needed more and better training, something Thorson and others eagerly signed up for. They needed a society that could accept paycheques from the government if they wanted to keep getting brought on as statutory hires. They needed more equipment. And they needed to become more professionalized. After seeing how much the organization had fizzled after the 2021 fires, Jessup was committed to keeping the home fires burning through the winter.

He recruited Peterson for his background in business and to help create a society. Then they convened a board of directors and started hashing out a formal structure for their organization. They called it the Knutsford Community Response Society, and the fire team itself was dubbed KIRT—the Knutsford Initial Response Team. They also doubled down on training, with a focus on their incident command and radio communication skills. Most people can figure out how to turn on a pump and point a nozzle at a fire—putting the wet stuff on the red stuff, as firefighters often say—but doing so in a coordinated, tactically useful way takes far more planning.

In the summer of 2024, KIRT got several more chances to prove themselves. One evening a burning truck started a grass fire in a ditch that threatened nearby homes. KIRT volunteers were on scene almost immediately, and while they lacked the equipment and training to fight the vehicle fire itself, they kept the grass fire from spreading until BC Wildfire and Kamloops structural fire crews arrived to take over.

On another day, a transformer on a utility pole blew, causing a fire not far from Jessup's house. Again, KIRT volunteers deployed and were on scene far faster than workers from the power utility company, who arrived in a truck with no water and no means to fight the fire themselves. BC Wildfire showed up not long after, but by then KIRT volunteers had the fire largely contained.

Their biggest challenge came in midsummer. Another lightning strike into the densely packed forest that Thorson now knew to fear. "If that thing had taken off, it would have been Rossmoore Lake all over again," he said.

Once again, KIRT volunteers were the first ones on scene and set to work digging a hand guard through the forest floor around the fire. They got pumps running and strung out hoses around the fire's flanks to attack it from the side—textbook initial attack tactics. When a BC Wildfire Service crew arrived, Thorson said they were impressed with the work KIRT crews had done, and KIRT started the process of handing over responsibility to professionals. According to Thorson, the BCWS firefighters insisted on setting up their own pumps, which seemed odd given that KIRT volunteers already had an effective water delivery system running. Even so, several KIRT crew members assisted with the switch over, but when it was time to turn on the water, the BCWS firefighters couldn't get their pumps to run.

"Eventually, I convinced them to just tie into my pump," Thorson said with a chuckle. Both crews worked side by side late into the night, getting the fire fully contained.

In the three years since the Rossmoore Lake fire, KIRT's work has been such a success that other municipalities often ask Jessup to give talks and workshops about how they did it. They've been profiled in local media, and—most importantly to Jessup—at around two-dozen volunteers, their ranks are finally beginning to feel stable. KIRT has been such a successful model for community-led volunteer wildfire fighting that it helped inform the creation of the BC Wildfire Service's Cooperative Community Wildfire Response (CCWR) program in 2024. The program is a set of formal guidelines that communities can follow to create formal societies like KIRT if they want to be included in helping fight nearby fires.

Other communities have followed a similar path, including a community fire brigade in Argenta, BC—a tiny, unincorporated community in the West Kootenay mountains that has been training and equipping its own firefighters for years. And though they've chosen to remain outside the formal CCWR program, they've still prioritized cooperation and trust building with the local wildfire zone officers. When a wildfire arrived on their doorstep in the summer of 2024, they were able to conduct an initial attack until the BC Wildfire Service arrived to help and took over control of the scene. When they did, the Argenta crew—under the leadership of Rik Valentine—had one final gift for the young incident commander.

Valentine had spent years carefully mapping every single viable water source in the community, from tiny creeks and culverts to stand pipes and small, almost invisible ponds. He used that knowledge to build an interactive, GPS-enabled digital map of every single one, which he handed over to the BC Wildfire crew, to the astonishment of the crew leader. Because wildland firefighters usually don't arrive at fires with trucks full of water, getting water-source mapping of that high quality on a remote fire is almost unheard of and hugely valuable. More than anything, the success of societies like KIRT, and the BC Wildfire Service's willingness to play ball with them, points to a larger more important shift away from a government that lectures and orders to one that's willing to cooperate.

Public demand for a better system to allow locals to fight fires had been growing for years and reached a crescendo after the debacle in Scotch Creek. After the CCWR was announced in early 2024, it was met with fanfare from some regions and disdain from others. In the North Shuswap, where calls for community involvement in firefighting were loudest, the regional district government rejected the CCWR program. The official reason was twofold: First, the district said, the costs of the program were too high, in particular the bureaucratic overhead that small municipalities would have to take on to get societies registered and administer the program. Second, North Shuswap residents were concerned they wouldn't be considered full partners in firefighting. "Our community is snake bitten by what has happened," Columbia Shuswap Regional District Director Jay Simpson told a local newspaper at the time. "This program doesn't cut it."

Beneath Simpson's official reasons for the rejection, I think there is a deeper truth. After what happened in 2023, the North Shuswap just didn't trust the government, and the government likely felt the same about the region.

That's the real lesson here. Underpinning all the work KIRT has been able to do in Knutsford is a foundation of trust, one that requires some give and take on both sides. In the North Shuswap that trust was broken early on, and neither side was able to repair it. On the Rossmoore Lake fire, Jessup's volunteers agreed to take an initially subordinate position on the faith that the fire service would respect them and let them stay. As Jessup said, they made a decision to trust and follow the BC Wildfire Service's lead, and in return they slowly earned the service's trust to do more complex and important tasks. "We played nice in the sandbox," Jessup said. It sounds simple—almost trite—but it works. Especially when the consequences of going the other way can be deep and long lasting.

If there is one thing I've learned in my years-long effort to better understand wildfires, it's that our time-tested ways of managing them are no longer enough. We need to completely rethink our relationships with fire. We need to put power, agency and money back in the hands of local communities. We need to devote more money to research and prevention, and we need to listen to what front-line firefighters are trying to tell us. Yes, we need more water bombers, but they and thousands more firefighters are not a solution. They are a necessary stop-gap to buy us time to flatten the curve and help restore a healthy balance to our forest landscapes and, ultimately, solve the existential threat of climate change itself.

There are clear answers to this crisis, but not necessarily easy ones. For any of this to work, we have to accept our current reality—one that experts tried to warn us about for decades—and we need to work together. Not everyone will have Terry Jessup's background, Landon Shepherd's experience, Lori Daniels's scientific insights, or Amy Cardinal Christianson's depth of history and culture. We don't all need to learn how to run Mark 3 pumps, dig hand guard or read Prometheus models. Some of us just need to clean out our rain gutters or trade our cedar privacy hedges for azaleas. But we had better learn how to do it together, and fast.

EPILOGUE

OUR FORESTS AND OUR FIRES ARE CHANGING

"Is that a fire?" my wife asked, squinting through the windshield at the mountainside north of Squamish. It was just after 4 on a warm June afternoon in 2025, and we were about to pull off the highway towards our friends' house nearby. We'd agreed to look after their two gregarious Bernedoodles while they were away on vacation.

Squamish sits about an hour's drive north of Vancouver, along the edges of Howe Sound. It's nestled in an estuary at the mouth of the river for which it's named, surrounded by mountains covered with the same wet hemlock, cedar and fir forests as Cypress, Grouse and Seymour mountains in North Vancouver. It's not the kind of place one expects to find a wildfire.

"Nah, I'm sure it's not. It's probably nothing," I replied, somewhat dismissively, as I eyed the wisps of grey and white floating over a subdivision about a kilometre ahead of us. When you chase wildfires as often as I do, every smokestack, campfire plume or low-hanging cloud has a tendency to trigger your "Ah-ha! I found it" instinct. Arriving at a wildfire just as it's starting is in some ways the perfect goal, and one that's almost impossible to achieve. It's what initial attack teams train for obsessively. Surely, I wasn't seeing a fire start before my very eyes.

But as we sat waiting for the traffic lights to change, those tendrils of smoke became a column. "Actually, maybe let's just go see," I said. We drove closer until I spotted a Squamish Fire Rescue truck parked on the shoulder of the road and a firefighter describing the blaze into a radio. As my wife turned the car around in a nearby subdivision, I messaged my editor at Reuters Pictures. A wildfire in the middle of Squamish is so unusual that even he initially assumed I was talking about something much more pedestrian, a dumpster fire perhaps. No, I assured him, this was most definitely a wildfire. I had managed to stumble upon it just as its flames began reaching for the tree canopy. Okay, he said, finally understanding. Time to get to work. After a quick stop at our friends' house to drop off my wife, I hauled on my Nomex and drove back towards the evacuate.

By the time I arrived, it was already gathering strength, burning at rank 2 and approaching rank 3. A small collection of onlookers had also parked on the highway shoulder, staring up at the fire now climbing the rocky hillside. Only three hundred metres away were the tightly packed homes on Tantalus Road as well as the nearby Skyridge townhouse complex. I pulled into a driveway just off the highway and found Gurmit Khattra and his family packing their vehicles, preparing to flee.

Khattra's house sat at the base of a steep cliff, only half a kilometre from where the fire had started. Prevailing inflow winds pushing north up the Howe Sound valley from the ocean were driving the fire across a ridge right above their property. I could already hear burning debris rolling down the steep slopes towards the house. I asked if there was anything they needed, but Khattra replied they were okay. They didn't seem particularly afraid, just cautious. Once the vehicles were loaded, the family pulled out chairs and sat on their wide, green lawn to watch the fire's progression, ready to flee if they had to. With the highway only a few metres away, it was an almost perfect vantage point to watch the fire's growth in relative safety.

I photographed them for a few minutes then drove back down the highway and circled around to the top of the Skyridge subdivision, south of the fire. There I found more than a dozen residents standing in the street, watching the show. By now, helicopters had been dispatched to the fire, and a BC Wildfire crew had hiked down a nearby mountain bike trail, working to set up water delivery. I heard the distinctive clatter as a Mark 3 wildfire pump roared to life somewhere near a creek below the houses. As I stood among the neighbours watching helicopters drop water from the sky, I realized no one else appeared to be preparing—like Khattra's family had—to evacuate. Perhaps they were confident in the wind direction, or perhaps they simply didn't realize the risk, and there were no officials around I could see to tell them. "You might want to start packing a few things," I said to a man filming the helicopters on his phone beside me. He just shrugged. I made some more photos from this new high vantage point then decided to head back to Khattra's house. The fire was already wrapping around the mountainside, out of view.

By the time I returned, the fire had moved a few hundred metres closer, and their yard was filling with smoke. I could hear the blaze crackling in the trees and occasionally the sound of a tree crashing down. Still, Khattra seemed unfazed. The winds were pushing the fire slowly towards them, but the steep slope behind their house gave the property a distinct advantage. Fire loves to burn uphill, and as long as the winds remained relatively light, chances were the fire would burn its way across the cliffs behind the house and away towards the north without reaching their home. But gravity was still a problem, and occasional burning debris kept rolling downhill, slowly spreading the fire closer.

As we watched the fire, a truck full of firefighters from Squamish Fire Rescue arrived and started taking stock of the scene. The house had decent defensible space. The wide yard wrapped about thirty metres around the home's southern and western sides, plenty of room to station large pumper trucks if they needed to. The eastern and northern edges of the property were much closer to the forest but still had enough space for firefighters to work. The crew started mapping out locations for sprinklers to help defend the home.

Suddenly, a small propeller-driven bird dog aircraft roared overhead, with loudspeakers blasting a warbling siren sound. A wisp of smoke trailed in a straight line behind the bird dog, marking a target for the tanker hot on its heels. Before long, the jet-engine tanker appeared, zooming through the smoke over the cliffs above Khattra's house and dropping a burst of bright red fire retardant in a broad paintbrush stroke across the treetops. As evening settled over the valley, the municipality issued evacuation alerts for one hundred properties near the blaze.

By the time the sun set, it was officially named the Dryden Creek wildfire. It was labelled "human caused," largely because there had been no lightning. Its birth, next to a popular mountain bike trail just after 4 p.m. that day, was also suspicious. It was discovered moments after it sparked by a local on the trail who—after calling 911—ran to a nearby house, grabbed a fire extinguisher and tried to put it out. But the fledgling fire had already grown too big.

It had ignited in a decadent stand of coastal hemlock, cedar and fir. Conventional wisdom and the experience of decades suggest that such fires are rare, and relatively low risk. The coastal fire zone doesn't get the kind of town-erasing megafires of BC's central interior or northern Canada's flammable boreal forest.

At least, it didn't use to. Complacent coastal residents sometimes point to a research paper from 2006 published in the *BC Journal of Ecosystems and Management* that detailed relative fire risks across British Columbia's coastal hemlock forests. These are the famously wet temperate rainforests of the Pacific Northwest. Locals lovingly drop the *S*, referring to it as the Wet Coast. It's the home of banana slugs and marbled murrelets, where Wade Simmons and Andrew Shandro pioneered free ride mountain biking in the 1990s. Surely the Wet Coast is still a haven from big fires, many hope.

That 2006 paper found that, historically, fires in "wet temperate coastal rainforests" were "very infrequent" and, when they did occur, were often low- to medium-severity fires, not the high-severity stand-replacing burns typically seen in dryer ecosystems. That matched the conventional wisdom of long-time local residents that fires were not something to worry about in places better known for their moss and mud.

Lori Daniels, the wildfire ecology researcher, knows that 2006 paper well. She and Bob Gray wrote it. "I can't tell you the number of people who still tell me 'Oh, we don't get fires here,'" Daniels later told me, with just a hint of exasperation in her voice. "Well, I can tell you, while it was true twenty years ago, it's not true today."

And just like the modern fires that now routinely outrun Prometheus, Daniels says the changes to our landscapes are happening far faster than even our best modelling predicted they would. "We're fifty years ahead of schedule," she said. Climate change has moved up the clock.

For a fire in early June, in a historically wet region not prone to wildfires, the Dryden Creek fire caught a lot of people off guard, including firefighters. Several told me they were shocked to see it triple in size so quickly. Had it started a few weeks later in even drier conditions at the height of wildfire season, there may not have been air tankers and wildland crews standing by to attack it. If there had been outflow winds that day pushing the fire towards town instead of inflow winds blowing it away, it could easily have showered embers onto those homes in the Skyridge neighbourhood, where residents had stood casually filming the helicopter water drops.

The morning after the fire started, one woman in a nearby subdivision woke up to find scorch marks on her front doormat, little pockets of black where embers had landed and smouldered but, thankfully, not ignited.

The Dryden Creek wildfire burned for thirty-eight days and grew to sixty hectares in size, but it didn't destroy any buildings and was ultimately declared out on July 17, 2025. Squamish had gotten lucky, just like so many Canadian towns and cities before it. Over the years, the more people I spoke to, the more burned homes and blasted cars and blackened forests I saw, the more this idea of luck began to stick in my head.

Canadians have not yet been confronted by a truly catastrophic wildfire. Don't get me wrong, thousands of people made homeless, entire communities left languishing under evacuation orders for months, lifetimes of memories gone up in flames—all of it is awful and traumatizing. Every wildfire death is a tragedy. But hundreds of destroyed homes is not the same as hundreds of dead people. We have not had our Lāhainā, our Los Angeles, our Black Saturday. Not yet, but we are living on borrowed time. As a former firefighter once told me, wildfires are inevitable. Loss isn't. How long it takes us to learn that lesson, and how many homes or lives it costs in the process, is up to us.

AS SQUAMISH'S BRUSH WITH FLAMES HELPS UNDERSCORE, OUR forests are changing rapidly. The way we deal with wildfires needs to change with them. Despite how badly fire services across Canada have struggled in recent years, those challenges are not because firefighters are bad at their jobs.

When towns and cities burn, it's not because firefighters are failing us. It's because we are failing them, with every expert recommendation we ignore, with every pay raise we don't give them, with every needle-clogged gutter we fail to clean out. We expect firefighters to hold the line for months at a time amid seasons their services were not built to face, against fires that can now easily shrug off any attempts to contain them.

The changes in our forests aren't just contained to Western Canada. The risks are growing across the country. In 2025, Newfoundland's Kingston wildfire, which destroyed nearly two hundred homes, burned until October. That same month more than 350 people were forced from their homes by a wildfire in Nova Scotia's Kings County. That year fires were still popping up in the forests of Ontario's cottage country despite it being nearly Halloween.

When I go to visit my mom at her home near Algonquin Park's western border, I typically fly into Toronto and drive three hours north, leaving behind the leafy Carolinian forests of southern Ontario. Around Gravenhurst, the sugar maple, red oak and ash trees begin to mix with towering white pines, spruce and balsam fir. In the fall it is especially beautiful, the hillsides swathed in orange, gold and red as the leaves change. I've made this journey so many times I could almost drive it in my sleep.

Eventually, I turn off the highway and I pass my old elementary school, where I used to play with my friends in the forest beyond the schoolyard at recess. We would collect stacks of fallen logs and heaps of downed pine boughs to build forts among the ranks of replanted conifers. It turns out that one- and ten-hour fuels woven together make great child-sized battlements.

These days, when I finally turn onto the dead-end dirt road that traces the lake's shoreline to the house I grew up in, I look at the forest and see not just trees, but fuel. As I pass the pine tree plantation, I imagine it shooting thirty-metre-high flames into the sky barely half a kilometre from my mom's front door or dropping embers into the tomato planters on her wide wooden deck. A place I never expected to see wildfire in my lifetime could easily burn today, and the prospect scares me.

But I also now understand that fear is part of the problem. It's what drove one hundred years of forest policy that tried and failed to eliminate wildfire all together. The impacts of that mismanagement took a century to manifest, but they're here now and they're not going away. The amount of work ahead of us is truly daunting. Millions of hectares of forest to thin, miles of red tape to slash, an unfathomable number of cedar shingles and juniper hedges to get rid of.

The fear that wildfires inspire is understandable, but we have to find a way to move beyond it. We have to stop treating wildfires like the monster in the deep, dark forest, something we're too afraid to look at until it catches us by surprise. We have to stop treating wildland firefighters like unskilled labourers working summer jobs. We must recognize them as the highly skilled year-round emergency responders that they are and recognize the level of resources they require. Most critically, we have to learn how to live with wildfire again, respecting both its awesome destructive power *and also* its vital role in sustainable, healthy landscapes. This, more than anything, is what will give our firefighters a fighting chance. If we don't do it, more of them will die battling wildfires they shouldn't have had to.

As I wrote these pages in October 2025, I resolved to do something that I—like too many people—had been putting off for years. I arranged for a FireSmart assessment of my mom's property at the lake. They're offered for free in most provinces.

An expert from FireSmart Canada would come and evaluate the risks, like the maple branches that lean close over the deck or the dense thickets of spruce and cedar trees downslope near the water. Then my mom would get a list of things she could do to make the property safer, like ensuring all the vents are covered with ember-proof mesh, or moving those tomato planters off the deck. Maybe she wouldn't need the assessment. Maybe the fire I feared would never come. But it was something concrete I could do, and it would help me sleep better in the smoky summer nights of our future.

ACKNOWLEDGEMENTS

This book would not have been possible without the patience and generosity of a huge number of people, many of them firefighters. I owe a debt of thanks to everyone who agreed to speak with me. In many cases, your words and experiences do not appear directly in these pages, but they were an enormous help in shaping my understanding of this crisis.

Specifically, thank you to Landon Shepherd, Matthew Conte, Don Smith, Colton Boutin, Ben Bartlett, Ty Barrett, Roy Phillips, Hinton fire chief Jim Smith, Riel Allain and many more who can't be named. You know who you are, and your commitment to service and to your fellow firefighters is remarkable.

In the world of wildfire science, I'm forever indebted to Lori Daniels, Amy Cardinal Christianson, Bob Gray, Mathiew Bourbonnais, Mike Morrow and Colleen Ross for helping me get my head around all the mind-blowing numbers.

To the community volunteers, both sanctioned and otherwise, thank you for trusting me with your perspectives and—in some cases—allowing me to document your work despite the potential legal risks of doing so. Thank you to Terry Jessup, Karl and Lea Thorson, Gord Peterson, Aron Wiens, and Craig Palmer in Knutsford, Ryan Cawkwell and Kody Kruesel in Monte Lake, and Ron Hamilton, Dave Dyck and Karl Bischoff in Scotch Creek, as well as all the residents of Woolford Point and Dorian Bay for so carefully documenting what you went through.

During my reporting on the Adams Lake wildfire, thank you to Mark Healey and Forrest Tower for granting me a rare opportunity to ask uncomfortable questions. Ben Boghean's explanation of fire behaviour analysis and the basics of fire modelling was also extremely helpful.

Thank you to Jim Cooperman for sharing the wealth of information and community testimonies you uncovered in your quest for public accountability.

To the evacuees, thank you to Claire Brookes and her husband Stephen Purcell, Brett Ireland, and Zoe Share for helping me understand the harrowing moments you lived through.

An extra-large thank you to Ollie Williams for speaking with me, and to the whole team at Cabin Radio for their incredibly deep coverage of the Northwest Territories fires and the Yellowknife evacuation. Their body of work, which continues today, is an invaluable resource that I relied upon to reconstruct much of how those fires were handled.

During wildfire safety courses, veteran forester Doug MacLeod taught me how to gauge fire weather indices, recite LACES by heart, and read an IAP, lessons I rely on every time I set foot near a wildfire, and critical training that should be mandatory for every journalist covering wildfire in Canada.

To every wildland firefighter I've stood next to on a fire line, thank you for allowing me into your world. There are too many to name, but a special thanks to Adam Buchanan, Billy Stevens, Jake Murie, Steve Lozano, Tyler Moylan, Carson Long, Eli Seligman, Aaron Schumacher, Fletcher Yancey, Charlie Helton, Jacqueline Cowley, Ashley O'Neil, Aws Al-Mubarak and the firefighters of the Fraser unit crew, the Nicola Knights, Al Ritchie, Dylan David, Rhys Jobbitt, Paul Ciulini and the rest of the Princeton Sierras (and their belt buckles), and the Columbia unit crew.

Thank you to my copy editor Kate Unrau and fact checkers Sarah Berman and Eva Holland for saving me from many embarrassing typos and errors. Thank you to my agent Sam Haywood for helping me navigate the book writing world, and to my ever-patient wife, Ainslie, for putting up with me through all the late nights and endless debates this project has required.

While everyone listed above made this book possible, my editor Lauren McKeon not only convinced me I could—in fact—write a book the first place, she made it readable, cutting through thickets of confusing prose and pulling me out of sometimes very deep rabbit-holes. Thank you to the whole team at HarperCollins Canada, and to Lauren especially for guiding a first-time author with such grace.

Throughout my career I've relied on the guidance of many incredible editors who shaped my work and helped me grow. Thank you to Chris Helgren at Reuters Pictures for always supporting my wildfire work, and being a staunch ally in the fight to improve wildfire press access in Canada. Thank you to my first photo editor Mike Thomas for teaching me to f/8 and be there, and to investigative editor David Bruser for teaching me that the story is in the details.

I've been lucky to build a career as a journalist by venturing far from the asphalt to places many people never get to see. This is only possible because I had the incredible good fortune to grow up paddling lakes and rivers, and hiking mountains with my brother Jon and our parents, Bonnie and Doug. You instilled in me a deep respect for the backcountry and taught me the power of a good yarn. As my dad used to say, sitting around a campfire with friends is one of the oldest forms of social interaction. I'm grateful for the many hours I got to spend doing that with him, discussing everything from politics to labour rights to the benefits of hand-stitched gable-ended tarps. I would not be the journalist, or the person, I am without you.

ENDNOTES

INTRODUCTION

3 **"You've found the head of the fire"** Personal interview, July 11, 2023.
6 **"they were going to set part of the forest on fire, on purpose"** BC Wildfire Service, *2019 Ignition Operations Manual* (2019).
10 **"Pretty incredible, isn't it?"** This quote and all subsequent comments attributed to Jake Murie are taken from personal interviews and field notes, July 11 and 12, 2023.
11 **"fire-adapted"** L. D. Daniels, et al., "Deciphering the Complexity of Historical Fire Regimes: Diversity Among Forests of Western North America." *Ecological Studies* 231, no. 14 (December 2017).
11 Jasper historical photos, mountainlegacyproject.ca.
12 **"Wildfire behaviour experts call it fire debt"** Matthew Bourbonnais, UBC Professor and wildfire behaviour researcher, personal interview, June 2025.
13 **"Fire-generated winds had topped 180 kilometres per hour"** Landon Shepherd, Parks Canada incident commander during the Jasper wildfire, personal interview, July 2025.
13 **"so disfigured forestry experts struggled to determine their species"** Lori Daniels, UBC professor, wildfire behaviour researcher and Koerner Chair in Wildfire Coexistence, personal interview, July 2025.

CHAPTER 1: K21620

16 **"185,000 square kilometres that burned"** *Canada Report: 2023 Fire Season* (Canadian Interagency Forest Fire Centre, 2023). https://ciffc.ca/sites/default/files/2024-03/03.07.24_CIFFC_2023CanadaReport%20%281%29.pdf.

17 David M. Romps et al., "Projected increase in lightning strikes in the United States due to global warming," *Science* 346, no. 6211 (14 Nov 2014): 851–854, https://www.science.org/doi/10.1126/science.1259100.

17 **"average May–October temperature across Canada was 2.2 degrees higher than normal"** P. Jain, Q. E. Barber, S. W. Taylor, et al., "Drivers and Impacts of the Record-Breaking 2023 Wildfire Season in Canada," *Nat Communications* 15, (2024): 6764. https://doi.org/10.1038/s41467-024-51154-7.

18 **"an average of ten to fifteen per year"** (On the number of fire spread days). X. Wang, D. K. Thompson, G. A. Marshall, C. Tymstra, R. Carr, and M. D. Flannigan, "Increasing frequency of extreme fire weather in Canada with climate change." *Climatic Change* 130, no. 4 (2015): 573–586. doi:10.1007/s10584-015-1375-5.

18 **"The worst-case scenario is an increase of 400%"** X. Wang, D. K. Thompson, G. A. Marshall, C. Tymstra, R. Carr, and M. D. Flannigan, "Increasing frequency of extreme fire weather in Canada with climate change," *Climatic Change* 130, no. 4 (2015): 573–586. doi:10.1007/s10584-015-1375-5.

22 **"Nearly five million of us—about 12% of Canadians—already live in the WUI"** S. Enri, et al., "Exposure of the Canadian wildland–human interface and population to wildland fire, under current and future climate conditions," *Canadian Journal of Forestry Research* 51 (2021). https://cdnsciencepub.com/doi/pdf/10.1139/cjfr-2020-0422.

24 Unless otherwise indicated, information describing IMT 5's handling of the Adams Complex wildfires, including the Rossmoore Lake, Bush Creek and Adams Lake fires, is drawn from internal government documents I obtained through BC's Freedom of Information process, including internal government communications, logbook notes, safety briefing documents and a detailed Facilitated Learning Analysis document that investigated how the events described unfolded.

26 Information describing how the wildfire unfolded near Dorian Bay and Woolford Point is drawn from contemporaneous meeting notes and records kept by local residents and shared with me during the reporting process.

29 **"just as the BC Wildfire Service's fire behaviour modelling predicted it would"** BCWS, "10-day growth projection, Adams Lake Wildfire, August 1-10, 2023." https://www.csrd.bc.ca/DocumentCenter/View/3579/Lower-East-Adams-Lake-Wildfire-K21620-Prom-Summary-Aug-1?bidId=.

30 **"The first of those planes arrived at 6:42 p.m."** According to local residents' contemporaneous notes and meeting minutes.

31 All scenes describing the Fraser unit crew's firefighting on August 4, 2023, near Dorian Bay and Woolford Point, are based on my on-the-ground reporting.

CHAPTER 2: IGNITION

38 Unless otherwise noted, information describing IMT 5's preparations for and execution of the planned ignition are drawn from internal government

records I obtained through the province's Freedom of Information process, including inspection reports, logbook entries, emails and text messages, and a detailed narrative description of events called a Facilitated Learning Analysis.

38 **"a last-ditch Hail Mary"** As described to me by a senior BC Wildfire Service officer.

47 **"Even if the backburn caused some damage, they reasoned, it had the potential to save far more"** Discussion of the "Trolley Problem" in *Adams Lake Wildfire Facilitated Learning Analysis* (BC Wildfire Service, June 26, 2024, p. 32).

49 **"80% of frontline crew members had less than twelve months' experience. Almost 60% of their crew leaders were in their first season as leaders"** BC General Employees Union internal pay-scale data, as provided to me by a source.

53 **"leaving only forty-three firefighters"** *Adams Complex Incident Action Plan* (BC Wildfire Service, August 16, 2023).

55 **"I've never done a large burn before," he told them. Despite these and other red flags, his crew was pressured into proceeding anyway"** Inspection report #202417752025A (WorkSafeBC).

CHAPTER 3: TRAPPED

59 Despite repeated attempts, I was unable to interview Ingham. She did not respond to phone messages or emails. All descriptions of her actions and feelings are based on internal government documents I obtained through the BC's Freedom of Information Act, including her detailed text message chain with Operations Chief Scott Reynolds.

59 Unless otherwise stated, descriptions of the BC Wildfire Service's handling of the planned ignition and its immediate aftermath are drawn from internal government documents I obtained from sources and through the province's Freedom of Information Act, including safety reports, logbook notes, text messages and emails, and an in-depth incident review report called a Facilitated Learning Analysis.

60 **"disbelief"** *Adams Lake Wildfire Facilitated Learning Analysis* (BC Wildfire Service, p. 4).

62 **"Shane, do not go in there"** Leanne Ingham to Shane Derhousoff, text messages obtained under BC's Freedom of Information Act.

64 **"It's blown over the powerlines and moving downslope"** Leanne Ingham to Operations Chief Scott Reynolds, text messages obtained under BC's Freedom of Information Act.

64 **"certain they'd just watched the trapped Brazilians die"** *Fire K21620 Fire Crew Entrapment Investigation—Full Report* (BC Wildfire Service, p. 6).

67 Unless otherwise noted, all information describing the events leading up to Devyn Gale's death and its aftermath are drawn from a copy of the *Facilitated Learning Analysis* (BC Wildfire Service) of the incident.

67 **"We're not fighting wildfires anymore"** Personal interview, summer 2023.

68 **"This led to the zone being understaffed and overworked"** *Facilitated Learning Analysis—Jordan River Wildfire Fatality* (BC Wildfire Service).

69 **"The incident up at the Donnie was a shock"** Unless otherwise noted, this and all subsequent quotes are taken from the *Facilitated Learning Analysis—Jordan River Wildfire Fatality* (BC Wildfire Service).

71 **"Blake"** The BC Wildfire Service gives pseudonyms to people referenced in FLA documents, to protect their identities and privacy. In the case of the FLA that investigated Devyn Gale's death, her family requested that her real name be used. I've chosen here to use the real name of her brother Nolan, who spoke at Devyn's memorial and has given public interviews about the day she died.

78 **"Oh fuck!"** Unless otherwise noted, this and all other quotes attributed to Dave Dyck are taken from a personal interview conducted in April 2025.

80 **"Celista Fire Chief Roy Phillips"** Unless otherwise noted, all quotes and information attributed to Roy Phillips are taken from several personal interviews conducted in April and May 2025.

CHAPTER 4: CHAOS, COURAGE AND CONSPIRACY

85 **"achievable objective"** Based on notes from Mark Healey's incident logbook (August 18, 2025).

87 **"We're here, in the fight"** Video of the firefighters' entrapment obtained from a source within the BC Wildfire Service.

88 **"The Bighorn and Firestalker unit crews"** As detailed in the *Adams Complex Incident Action Plan for August 18, 2025* (BC Wildlife Service).

88 **"Dean Acton and his son Mark"** Unless otherwise noted, information describing the actions of Dean and Mark Acton, as well as several over local residents who stayed behind after evacuations were ordered, is based on reporting by CBC News, as well as evidence and testimonials collected by Lee Creek resident Jim Cooperman. While I corresponded with Mark during the reporting of this book, neither he nor Dean agreed to an interview.

88 **"Twelve of these monster water cannons were arranged along a snaking three-kilometre line"** Contractor layout document detailing Scotch Creek structure protection plan obtained from a Scotch Creek source.

91 **"The fire now threatened to cut off the only road out for hundreds of people living east of Scotch Creek"** Information in this section describing the actions of structural firefighters like Roy Phillips and Ty Barrett, as well as fire behaviour in areas where they worked, is taken from interviews conducted in late 2024 and early 2025.

91 **"'You need to get your guys off of this side of the bridge and back to the other side,' he said"** According to Roy Phillips.

92 **"Radio communication across the region was either failing or jammed with traffic from many firefighters trying to relay messages to each other at once"** According to both Phillips and Barrett.

95 **"Firefighters who were there described scenes of absolute chaos"** *Adams Lake Facilitated Learning Analysis* (BC Wildfire Service, p. 56).

97 **"We don't put firefighters in front of rank six fire"** Mark Healey, personal interview, October 2023.

97 **"Contractors and other structure protection specialists who had been sheltering near the marina were ordered to evacuate at around 8:45 p.m."** Mike Graeme, "What it's like to flee a wildfire in BC," The Narwhal.ca (August 24, 2023).

97 **"At one point, the back of the local Home Depot caught fire. Locals ripped the siding off and doused the flames, likely saving the building"** According to local testimony gathered by Lee Creek resident Jim Cooperman.

98 **"inferno road"** *Adams Lake Facilitated Learning Analysis* (BC Wildfire Service, p. 40).

98 **"Healey—alone—turned east"** *Adams Lake Facilitated Learning Analysis* (BC Wildfire Service, p. 41).

CHAPTER 5: RADIO SILENCE

114 **"With only around 150 Type 1 firefighters in the whole territory, and 250 Type 2s"** Matthew McClearn, "Canada's wildland firefighters are stretched thin. Where are resources falling short, and can they catch up?" *Globe and Mail* (Sept. 20, 2024).

114 **"NWT didn't have any highly-trained incident management teams capable of managing large, complex fires"** *After-Action Review, Northwest Territories 2023 Wildfire Season* (Transitional Solutions Inc., May 2025, p. 70).

114 **"Firefighters from different municipalities and jurisdictions sometimes wound up working at odds with each other"** *City of Yellowknife After Action Assessment: 2023 North Slave Complex Wildfires* (KPMG, p. 96).

114 **"Political interference was more cumbersome than operations, yet no one wanted to make decisions"** *After-Action Review, Northwest Territories 2023 Wildfire Season* (Transitional Solutions Inc., May 2025, p. 70).

115 **"What little predictive modelling NWT firefighters did have was sometimes either misunderstood or disregarded"** *After-Action Review, Northwest Territories 2023 Wildfire Season* (Transitional Solutions Inc., May 2025, p. 96).

119 **"Yellowknife is *not* under threat of wildfire"** *After-Action Review, Northwest Territories 2023 Wildfire Season* (Transitional Solutions Inc., May 2025, p. 98).

124 **"Some people would say, 'yes, it's definitely coming' and it was sort of like 'well, you said that last week'"** This, and all subsequent quotes attributed to Zoe Share are taken from personal interviews conducted in May 2025.

124 **"emails obtained by Cabin Radio reporter Emily Blake"** Emily Blake, "Yellowknife evacuation emails show confusion, lack of planning," CabinRadio.ca (May 14, 2025).

126 **"I would really start thinking carefully about an evacuation plan and imagining what that might be like, and that's something that they did not do in Fort McMurray literally until the last minute"** Ollie Williams's recollection of his interview with author John Vaillant.

127 **"Why am I still on this side of this wildfire?"** Unless otherwise noted, this and all other quotes attributed to Ollie Williams are taken from personal interviews conducted in late 2024 and early 2025.

129 **"we'll just figure it out"** Unless otherwise noted, this and all other quotes attributed to Claire Brookes are taken from personal interviews conducted in April 2025.

129 **"Starlink"** When the fire cut the fibre-optic internet connection, the territory lost its only physical infrastructure connecting it to the wider internet. Starlink units, the small, mobile satellite internet receiver system created by Elon Musk, became critical to the GNWT's ability to keep its operations centre running.

CHAPTER 6: EVACUATION

148 **"Everybody's truckin' up here. There's nobody standing around, and we're going to be ready for this thing"** Chris Greencorn, as quoted in "These People are Building Yellowknife's Fire Break," Cabin Radio (August 23, 2023). https://www.youtube.com/watch?v=-7F82xPRTUY&t=463s.

148 **"It's a level of cooperation among local contractors that I've never seen, and I've been here over thirty years"** Kenny Ruptash, as quoted in "These People are Building Yellowknife's Fire Break," Cabin Radio (August 23, 2023). https://www.youtube.com/watch?v=-7F82xPRTUY&t=463s.

151 **"City staff had done little training or preparation for an emergency of this scale, the report found"** City of Yellowknife After Action Assessment: 2023 North Slave Complex Wildfires (KPMG, p. 38).

151 **"the sudden effort to build massive fire breaks and sprinkler lines around the city diverted critical resources away from planning for the evacuation"** City of Yellowknife After Action Assessment: 2023 North Slave Complex Wildfires (KPMG, p. 66).

152 **"glaring holes in the GNWT's preparedness, including inadequate fire behaviour modelling, outdated data tools and a lack of appropriate training for fire behaviour specialists"** Northwest Territories 2023 Wildfire Response Review, Final Report (MNP, July 2024, p. 193).

152 **"bad coordination between all three levels of government"** Northwest Territories 2023 Wildfire Response Review, Final Report (MNP, July 2024, p. 5).

152 **"lacked enough trained staff for 'complex' operations like indirect attack"** Northwest Territories 2023 Wildfire Response Review, Final Report (MNP, July 2024, p. 191).

152 **"the territory did, in fact, have an emergency management plan all along. It had been created in 2018, and it 'should have worked'"** After-Action

Review, Northwest Territories 2023 Wildfire Season (Transitional Solutions Inc., May 2025, p. 53).

155 **"Given the NWT's small population, limited fiscal and human resources, and the infrequency of large-scale emergencies requiring sustained territorial-level response, a separate agency would be costly, duplicative, and difficult to staff"** Government of the Northwest Territories official statement, as quoted in Ollie Williams, "GNWT rejects key 2023 recommendation but accepts others," CabinRadio.ca (October 2, 2025).

CHAPTER 7: CITIES ON FIRE

158 **"wake-up call"** *The Silver Creek Fire Review* (Ombudsman of the Province of British Columbia, 1998, p. 9).

158 **"the conditions of 1998 were the most severe ever experienced"** *The Silver Creek Fire Review* (Ombudsman of the Province of British Columbia, 1998, p. 9).

158 **"Another report"** The Honorable Gary Filmon, *Firestorm 2003: Provincial Review* (2004). https://www2.gov.bc.ca/assets/gov/public-safety-and-emergency-services/wildfire-status/governance/bcws_firestormreport_2003.pdf.

159 **"another government report"** George Abbott and Chief Maureen Chapman, *Addressing the New Normal: 21st-Century Disaster Management in British Columbia* (2018).

159 **"Fourteen years after the Filmon report was completed, hundreds of large and small communities across British Columbia remain vulnerable to wildfire"** Abbott and Chapman, *Addressing the New Normal* (2018, p. 109).

163 **"A 2023 paper authored by Leona Shepherd and Mike Flannigan"** "Future wildfire spread days in Wells Gray Provincial Park," Thompson Rivers University, 2023, https://www.tru.ca/research/research-centres/wildfire-science/research-projects/Future_wildfire_spread_days_in_Wells_Gray_Provincial_Park.html.

169 **"Fire entered the communities almost exclusively via burning embers, rather than direct flame front, and quickly shifted from being a wildfire to a set of several urban fires"** *A wildland-urban post-fire case study: The Grouse Complex* (FP Innovations, 2023, p. 3).

173 **"the Adams Lake fire saw BUIs of more than 130"** *Adams Complex Incident Action Plan* (BC Wildfire Service, August 17, 2023).

173 **"the Northwest Territories that summer saw BUIs of 200"** Mike Westwick, GNWT Wildfire Information Officer, public fire update, August 2023.

174 Wendy Stueck, "Fighting Fire with Fire," *Globe and Mail* (May 28, 2023).

174 *Forest and Fire Management in BC: Toward Landscape Resilience* (British Columbia Forest Practices Board, June 2023). https://www.bcfpb.ca/release-publications/releases/forest-and-fire-management-in-bc-toward-landscape-resilience/

174 *Wildfire Season Summary* (BC Wildfire Service, April 2025).

174 New Jersey Department of Environmental Protection, "Deep Forest Fire Service Ramps Up Annual Prescribed Burning Program, Unveils New Jersey Wildfire Risk Assessment Portal," news release, February 27, 2024.

175 Nirmal Subedi, Bryan Bogdanski, and Brad Stennes, "Estimated Direct and Indirect Costs of Extreme Wildfires in Western Canada," Natural Resources Canada, 2005.

CHAPTER 8: CALL FOR AID

179 **"There is a lot of fire scar activity right up to about 1900 just before the park was created"** Personal interview, July 2025.

181 **"It would become known as the 'north fire'"** Unless otherwise noted, scenes describing the fire's behaviour and firefighter actions in Jasper are based on personal interviews with Jasper Fire Chief Matthew Conte, Jasper Deputy Fire Chief Don Smith, Parks Canada Deputy Incident Commander Landon Shepherd, Hinton Deputy Fire Chief Colton Boutin, Parks Canada wildland firefighter "Regan," and Jasper Brewing Co. founder Brett Ireland completed over the course of 2024 and 2025.

182 **"They should start flying tonight, they shouldn't wait till tomorrow"** According to Landon Shepherd, Parks Canada deputy incident commander, July 2025.

183 **"This highway is going to be closed within the hour"** As quoted in Connor O'Donovan, "Startling video shows first moments of the south Jasper wildfire," Weather Network, August 3, 2024.

184 **"Oh my God, they're closing the highway"** According to Colton Boutin, Hinton deputy fire chief, November 2024.

186 **"three separate lightning strikes at 7:05, 7:06 and 7:08 p.m."** *Jasper 2024 Wildfire Complex 2024: Fire behaviour, documentation, reconstruction, and analysis* (Canadian Forest Service, October 2025, p. 18).

187 **"There is no immediate threat to the town of Jasper"** Town of Jasper, public statement, July 22, 2024.

188 **"Less than two hours after they were emptied of tourists, both Wabasso and Whirlpool were completely burned over"** According to Parks Canada's official timeline of the Jasper wildfire.

CHAPTER 9: WE'RE NOT GOING HOME TODAY

193 Unless otherwise noted, scenes describing the fire's behaviour and firefighter actions in Jasper are based on personal interviews with Jasper Fire Chief Matthew Conte, Jasper Deputy Fire Chief Don Smith, Parks Canada Deputy Incident Commander Landon Shepherd, Hinton Deputy Fire Chief Colton Boutin, and Parks Canada wildland firefighter "Regan," completed over the course of 2024 and 2025.

195 **"wind speeds reaching more than two hundred kilometres per hour"** *Jasper 2024 Wildfire Complex 2024: Fire behaviour, documentation, reconstruction, and analysis* (Canadian Forest Service, October 2025, p. 52).

195 **"pilots' heads banged against their cockpit walls"** According to Landon Shepherd, Parks Canada deputy incident commander at the time.

195 **"If the winds died down they could come back"** According to Landon Shepherd, Parks Canada deputy incident commander at the time.

197 **"it wasn't long before hydrants in the Pacific Palisades and other higher-elevation neighbourhoods began to run dry"** According to Hannah Fry and Ian James, "After Palisades fire hydrants went dry, LAFD faced costly delays in getting more water," *LA Times* (October 11, 2025).

199 **"People do not understand what fire-smarting is"** Parks Canada wildland firefighter "Regan," personal interview, summer 2025.

201 **"Fire-generated winds reached 260 kilometres per hour"** *Jasper 2024 Wildfire Complex 2024: Fire behaviour, documentation, reconstruction, and analysis* (Canadian Forest Service, October 2025, p. 29).

203 **"Guys, we're not going home tonight. This thing is gonna be here today"** Colton Boutin, personal interview, November 2024.

CHAPTER 10: THE BATTLE FOR JASPER

206 **"It was like a volcano had gone off"** Unless otherwise noted, this and all subsequent quotes attributed to Ben Bartlett are drawn from a personal interview, November 2024.

206 **"We think it's a couple hours out, maybe an hour"** According to Colton Boutin, personal interview, November 2024.

207 **"Christine, we have to leave"** Landon Shepherd, as quoted in Matthew Scace, *Jasper on Fire: Five Days of Hell in a Rocky Mountain Paradise* (Sutherland House, 2025).

207 **"Right now?"** Christine Nadon, as quoted in Matthew Scace, *Jasper on Fire: Five Days of Hell in a Rocky Mountain Paradise* (Sutherland House, 2025).

208 **"The ignition chopper lifted off around 4:40 p.m."** *Jasper 2024 Wildfire Complex 2024: Fire behaviour, documentation, reconstruction, and analysis* (Canadian Forest Service, October 2025, p. 33).

209 **"It's too damn smoky"** According to Parks Canada wildland firefighter "Regan," personal interview, June 2025.

211 **"Get the hell out of there! It's gonna go right over you!"** According to Hinton Deputy Fire Chief Colton Boutin, personal interview, November 2024.

213 **"The column's collapsed, where are you?"** According to Parks Canada wildland firefighter "Regan," personal interview, June 2025.

215 **"Okay, I know how to start a saw"** Unless otherwise noted this, and all other quotes attributed to "Regan" are taken from a personal interview, June 2025.

219 **"at times topping 110 kilometres per hour"** *Jasper 2024 Wildfire Complex 2024: Fire behaviour, documentation, reconstruction, and analysis* (Canadian Forest Service, October 2025, p. 35).

221 **"it would take hours. It would start off, it would get bigger, it would burn for a good two hours and then it would finally collapse into the basement"** Unless otherwise noted, this and all other quotes attributed to Jasper Deputy Fire Chief Don Smith are taken from a personal interview, November 2024.

223 **"Fully involved"** Unless otherwise noted, this and all other quotes attributed to Jasper Fire Chief Matthew Conte are taken from a personal interview, November 2024.

CHAPTER 11: RESET

230 Roughly 7.5 kilometres long by 2.25 kilometres wide, measured conservatively with satellite imagery based on my own aerial photos.

232 Kristopher Liivam testimony, Standing Committee on Environment and Sustainable Development, Parliament of Canada, Ottawa (October 7, 2025).

234 **"Their system of fuel treatment areas around the town was more complete and extensive than any other Canadian city ever hit by a wildfire"** *Jasper 2024 Wildfire Complex 2024: Fire behaviour, documentation, reconstruction, and analysis* (Canadian Forest Service, October 2025, p. 54).

235 **"These are three-hundred-year-old Douglas firs that survived the valley burning wall to wall at least twice in the past 250 years"** Unless otherwise noted, this and all other quotes attributed to Parks Canada Incident Commander Landon Shepherd are taken from a personal interview, June 2025.

235 **"red phase"** This description of pine beetle impacts draws heavily on the work of Tristan Skretting, in her study "Forest fuel structure and loading along a gradient of grey-phase mountain pine beetle severity in Jasper National Park, Alberta, Canada," *Canadian Journal of Forest Research* (April 2025).

236 **"386,000 kilowatts of energy per metre"** *Jasper 2024 Wildfire Complex 2024: Fire behaviour, documentation, reconstruction, and analysis* (Canadian Forest Service, October 2025, p. 49).

236 **"seven times more energy than the nuclear bomb dropped on the Japanese city of Hiroshima"** Daniel Rosenfeld et.al. "The Chisholm firestorm: observed microstructure, precipitation and lightning activity of a pyro-Cb" *Atmospheric Chemistry and Physics Discussions,* October 2006.

237 **"It should set off alarm bells for everyone concerned about the impacts of climate change"** Lori Daniels, email, October 21, 2025.

239 **"Fundamentally, it does sort of precipitate this idea that fire is just bad, and that's not the case"** Mathieu Bourbonnais, personal interview, June 2025.

242 **"The land knew that we needed to restart."** Unless otherwise noted, this and all other quotes attributed to Ashley O'Neil are drawn from a personal interview, May 2025.

243 **"We sort of picture BC as wall-to-wall conifers"** Wildfire ecologist Bob Gray, personal interview, June 2025.
247 **"We want to be good stewards"** Colleen Ross, personal interview, May 2025.
250 **"I'm on a journey of reconnection with fire and have been really lucky to have a bunch of elders who've taught me and supported me"** Unless otherwise noted, this and all other quotes attributed to Amy Cardinal Christianson are taken from a personal interview, May 2025.
251 **"We are recklessly destroying the timber of Canada and there is scarcely a possibility of replacing it"** Sir John A. Macdonald, as quoted in Douglas Martin, "Canada's Wasted Woodlands," *New York Times* (August 28, 1983).
251 **"Conservation . . . involves preventing destruction by fire, and secondly, a judicious system of cutting the timber so as to retain for all time a continuous supply"** As quoted in T. G. Honer and Kenneth Johnstone, "Elihu Stewart and the Beginnings of Dominion Forestry," *The Forestry Chronicle* 60, no. 4 (1984): 225–230.
254 Lori Daniels et al., "The 2023 wildfires in British Columbia, Canada: Impacts, Drivers, and Transformations to Coexist with Wildfire," *Canadian Journal of Forest Research* (2025).

CHAPTER 12: COMMUNITIES AT THE READY

256 **"C'mon in!"** Unless otherwise noted, this and all other quotes attributed to Terry Jessup are taken from a personal interview, May 2025.
259 **"It could have been better"** Aron Wiens, personal interview, May 2025.
261 **"Oh shit"** Unless otherwise specified, this and all other quotes attributed to Karl Thorson are taken from a personal interview, May 2025.
262 **"Where do you take forty-five horses?"** Lea Thorson, personal interview, May 2025.
268 **"Right from the start, I think it was pretty important that there was no adversarial stuff going on. The people who decided to stay and fight the fire were respected"** Unless otherwise specified, this and all other quotes attributed to Gord Peterson are taken from a personal interview, May 2025.
275 **"Our community is snake bitten by what has happened"** Jay Simpson as quoted by Barb Brouwer, "Columbia Shuswap directors refuse downloading of public wildfire response training," *Vernon Morning Star* (February 22, 2025).

EPILOGUE: OUR FORESTS AND OUR FIRES ARE CHANGING

283 **"Well, I can tell you, while it was true twenty years ago, it's not true today"** Lori Daniels, personal interview, June 2025.
284 **"More than 100 Squamish properties now under evacuation alert,"** Global News, posted June 10, 2025, 5 min., 41 sec.